**W9-CIK-122**

THE WHISPER AND THE VISION

We were dreamers, dreaming gently, in the man-stifled town;
We yearned beyond the sky-line where the strange roads go down.
Came the Whisper, came the Vision, came the Power with the Need,
Till the Soul that is not man's soul was lent to us to lead.

Rudyard Kipling, *A Song of the English*

# the Whisper & the Vision

## THE VOYAGES OF THE ASTRONOMERS

Donald Fernie

CLARKE, IRWIN & COMPANY LIMITED,
TORONTO/VANCOUVER

Fernie, Donald, 1933-
  The whisper and the vision

ISBN 0-7720-1090 0


1. Astronomers.   2. Astronomy - History.
I. Title.


QB35.F47        520'.92'2        C76-017174-2

Published simultaneously in the United States by
Books Canada Inc., 33 East Tupper Street, Buf-
falo, New York 14203, and in the United Kingdom by Books Canada Limited, 1 Bedford Road,
London N2.

1 2 3 4 5 JD 80 79 78 77 76

Printed in Canada

# CONTENTS

# LIST OF ILLUSTRATIONS

# PREFACE

This is not a book about astronomy. It is a book about people: some of the people from the past who made astronomy what it is today. Its theme is the experiences and adventures of travelling astronomers in the eighteenth and nineteenth centuries, men who set out across the earth, and who at times underwent astonishing hardships, in the pursuit of their science. It might have been a history of astronomy, using the practitioners to illuminate their subject, but it is not that either. For the trouble with setting out to write a proper history is that one must be fair, or at least balanced, and that means taking a great deal of the colourless with the colourful. That, of course, has its proper place, but this is not it. Nor, for that matter, have I chosen my subjects only from the giants of astronomy; many of the most important astronomers do not appear here, and some who do were relatively minor figures.

This, then, is not a scholarly book. It is an informal, light-hearted stroll down the corridors of astronomical history, peeking in at doors chosen by no criterion but my whim. If it must have a purpose, then that purpose is no more than to bring astronomers down to human scale and show them in the context of the times in which they lived. Professional astronomers are often only too aware of how the public views them: a mixture of awe and pity and sometimes even derision for these earnest intellectuals far removed from reality, forever debating events a million light-years away. I hope these pages will modify that view and show many astronomers to be rather ordinary people very much concerned with the realities of life.

Lastly, I must report that none of the material in this book comes directly from original manuscript sources, and I therefore owe a considerable debt to authors who have gone before me. In particular, I would like to mention Helen Hogg, who first drew attention to the journals of William Wales, and who translated the remarkable account of Le Gentil's voyages. It is also a pleasure to thank David Evans, without whom we would know very little of the adventures of Maclear and Herschel, and to whom I am particularly indebted for permission to use a small amount of his unpublished material on Maclear. These and other sources are cited in the bibliography.

*Toronto*
*July, 1976*

DONALD FERNIE

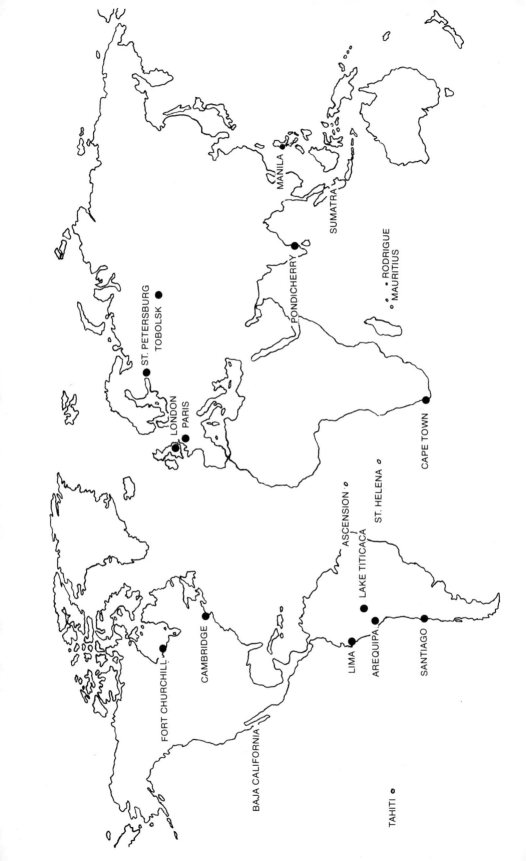

MANILA

SUMATRA

PONDICHERRY

RODRIGUE
· MAURITIUS

ST. PETERSBURG ●
TOBOLSK ●

LONDON
PARIS

CAPE TOWN

ASCENSION ○

ST. HELENA ○

LAKE TITICACA

FORT CHURCHILL

CAMBRIDGE

LIMA

AREQUIPA

SANTIAGO

BAJA CALIFORNIA

TAHITI ○

TRANSITS AND TRIBULATIONS

Late September. Early Spring. And the silence, my God, the silence, beating gloriously on one's ears. Only the sibilant breeze to break it, whispering over the ridge, stirring the sparse stunted shrub, passing on to the harsh rock and sand below. Nothing moves.

Sitting there in the high thin air, shirtless in the afternoon's warm sunshine, I feel utterly content. Overhead the cloudless sky, a deeper blue than ever, for I am almost 8,000 feet above sea level. Below me the foothills of the Andes falling away in endless folds of gold and purple across the Atacama Desert to the blue line of the Pacific fifty miles away. Behind me, beyond the ridge, the mountains rising one upon the other to the snowcapped spine of the Andes, another fifty miles away on the Argentine border. The white hulk of 21,000-foot Cerro del Torro.

I sit back and reflect. To the left, twenty miles due south, the little white domes of the European Southern Observatory, clustered on another ridge. Further away, beyond the scope of even this vantage point, more white domes of the Cerro Tololo InterAmerican Observatory, while stretched down this ridge on which I am sitting, the domes of the Carnegie Institute and our own telescope.

We are the new peripatetic astronomers, jetting down to Chile from Canada and the States and Europe, stopping over to enjoy the luxuries of Santiago, driving up the mountains in modern trucks to where good food and comfortable dormitories await us. Around us the gleaming telescopes and accessories, their electronic complexities watched over by technicians and resident engineers. A week, two weeks, to obtain our data under the brilliant skies, and then jetting off home again for the months of analysis. And elsewhere other such groups, in Australia and Africa, Spain and Hawaii, wherever the skies are calm and clear. For some of us it may even be this fortnight in Australia, the next in Chile, and home to Europe by the end of the month, casually spanning the world.

Two condors, forsaking the high snows in search of food, glide across the ridge, their giant eight-foot wings spread gracefully and effortlessly as they float against the updrafts. I get a brief baleful glance from an ugly pink head, and they pass down the valley.

I yawn and stretch, trying to concentrate on a decision as to what sort of sandwiches I should order for midnight lunch tonight. What, I wonder, would Solon and Marshall Bailey have made of all this? Almost a hundred years have passed since they scaled some of these selfsame peaks

I see before me now. But then it was a matter of foot-slogging, and you needed guides and packmules; you weren't always sure where your next food and drink would come from, and above all you hoped to God you wouldn't break a leg out here in the mountainous desert.

It is getting late, and already the evening fog is starting to creep in over La Serena on the coast. As night comes, the fog will sidle up the long valleys, and when in the pre-dawn dusk I again come down the path from the telescope, there will be spread before me a view as stunning as the one I now see: a great grey ocean of fog thousands of feet below, billowing silently as though in supplication to the implacable peaks of the Andes, already roseate in the early dawn.

I fold my chair and with a sigh start back up the path to the cottage. What would Solon Bailey have had on his sandwiches?

Astronomers, by and large, are a travelling lot. Rather surprising, you might think. If the earth is such a small speck in space, what does it matter from where on it one views the cosmos? And it's true that the perspective of the distant stars and galaxies does not change with position on the earth. But there are nevertheless some very valid reasons for moving astronomers around the earth.

The most obvious, I suppose, is going in search of clear skies. This is not quite as simple as it sounds, because much of what the average person would call good weather is in fact not much good to the astronomer. For instance, the islands of the Caribbean are generally regarded as havens of sunshine by tourists, yet you will find that there are almost always clouds in the region. True, they are often broken clouds, or high thin clouds, which allow plenty of tanning of pale Canadian skins; but nevertheless they are anathema to the astronomer. Modern astronomy, you see, is not so much a matter of *looking* at things through a telescope as of *measuring* things through a telescope. It is easy enough now to measure the brightness of a star no brighter than a candle in London viewed from New York, but obviously even the thinnest of clouds will ruin the result.

A more subtle requirement is steadiness of the surrounding atmosphere. Any shimmering, so easily shown up in a high-power telescope, will spoil the astronomer's efforts. And so some places which have clear skies, such as the Sahara Desert of North Africa, are no good to the astronomer either.

Lately a new twist has been added. Astronomers are now more and more working outside the limits of visible light, studying the stars by, for example, infra-red light. Since infra-red light is easily absorbed by water vapour, the new sites must also have very low humidity.

With all these requirements—and several others as well—to be met, it isn't too surprising that there are relatively few places on earth which astronomers find really satisfactory. And that is why they are prepared to travel thousands of miles to out-of-the-way places like northern Chile to make their observations.

But the idea of travelling great distances to make routine observations is a recent innovation. In the past long-distance travelling for this purpose was, in varying degrees, too hazardous, too tedious, and too time-consuming to be practicable. Only with the coming of modern air travel have astronomers begun flitting around on an almost day-to-day basis. Historically, astronomers travelled for different reasons.

One of these arose from the simple fact that stars in the far southern skies are not visible from the latitudes of Europe and North America. So if an astronomer wished to be complete in his celestial surveys there was nothing for it but to travel to some country in the southern hemisphere to continue his work. But more than that was often involved. As the maritime nations extended the operations of their fleets during the seventeenth, eighteenth and nineteenth centuries, the need for navigational information in the southern hemisphere became more and more pressing. This led to the setting up of major, permanent observatories in countries such as Australia, South Africa, and parts of South America. These observatories were usually staffed by men from the mother countries, and since few of them found life in the colonies to their liking, there was soon a steady trickle of astronomers going out to or returning from their outposts.

There were good reasons for astronomers to travel not only north and south, but also east and west. A once-popular song has it that "when it's night-time in Italy, it's Wednesday over here", which, I suppose, is sometimes true. Certainly, though, when it's daylight in one country it will be night in some other country, and if astronomers become interested in some rapidly changing event in the heavens they will want the co-operation of their fellows east and west of them to provide continuity in the observations. As dawn comes to one astronomer, who must then cease work, darkness will fall for another, who can then begin work. And so there has sometimes been travel east and west by astronomers to establish such networks of observers.

But perhaps the greatest amount of travel, historically, was undertaken by astronomers bent on single events. And if the first link between these men was their astronomy, the second was surely their courage and tenacity. The strength in the face of adversity shown by these peripatetic astronomers of the past is a theme which runs through all their stories, a theme that raises the fascinating question of what dreams or visions or psychological needs (or perhaps just cheerful willingness?) drove these people to achieve what they did. The cost to some, as we shall see, was their very lives; one man spent twelve years travelling more than thirty thousand miles to observe two events lasting only a few hours, and was defeated on both occasions. What motivated such sacrifices is a question to which we shall return.

Almost the first of the great astronomical travels came in the eighteenth century through an event known as a transit of Venus, and it with attempts to record these transits that the rest of this chapter is concerned. Venus is a planet which is closer to the sun than is the earth, and, as Venus and the earth revolve about the sun, there are occasions when the three objects are lined up with Venus in the middle. At such a time Venus will appear to an earthbound observer as a small black blob projected against the fiery disk of the sun, slowly travelling in the course of a few hours from one edge of the sun's disk across to the other. This transit of Venus is a rare event. Transits come in pairs, eight years apart, and with something over a hundred years between pairs. Thus, for instance, there were transits in 1761 and 1769, again in 1874 and 1882, and the next are due only in 2004 and 2012.

Historically, the importance of the transits of Venus lay in the fact that they were once thought to provide the most accurate means for determining the distance between the earth and the sun. Two astronomers, widely separated on the earth, will each see Venus projected against slightly different parts of the sun's disk, and so will find the duration of the transit to be different. If the positions of the observers on the earth are known (this is important), and if the duration of the transit is timed by each of them, then it is possible to calculate the distance between the earth and the sun.

Although in principle only two observers in different parts of the earth are needed, it would never be safe in practice to rely on the weather being

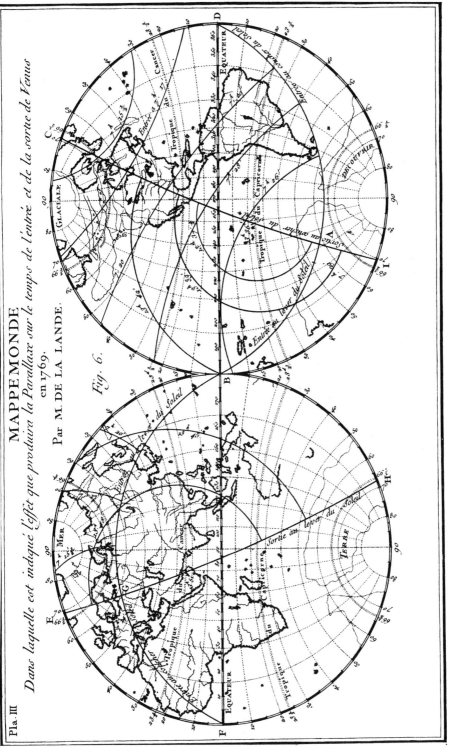

Mappemonde *for the Transit of Venus in 1769 (taken from Harry Woolf,* Transits of Venus, *(c) 1959 by Princeton University Press, and reprinted by permission of Princeton University Press)*

clear at the critical moment at any one site. Accordingly, numerous expeditions to many different sites were always sent out, each equipped with sufficient instruments to measure the time when the transit began and the time when it ended, and also the latitude and longitude of the observing site. Additionally, in case of sickness or accident, there were usually at least two people capable of making the observations on each expedition.

But was it worth the months, indeed years, of preparation, the great expense, the immense efforts of the astronomers themselves, merely to measure the distance between the sun and the earth? The answer is yes, because until quite recently finding the sun's distance was much more than an end in itself. It lay at the very root of astronomy. This came about because our knowledge of distances in the universe is built up on a step-by-step basis. When the distance of the sun is accurately known, it serves as a basis for finding the distances of the nearer stars. And when these are known they serve as calibrators of other methods for finding the distances of more remote stars. Knowing these the astronomer can proceed to find the distances of galaxies beyond the Milky Way, which in turn offer a means of measuring the distances to galaxies yet further away. And so on to the limits of the observable universe. Ultimately, our knowledge of the very kind of universe in which we live rests on these distance scales. Through them we can confront such questions as What is the age of the universe? Will it expand for ever? So yes; although in the eighteenth century much of this could not be foreseen, finding the distance to the sun was already crucial to astronomy.

Yet this was not easy. Aristarchus, one of the early Greeks, had a stab at it as long ago as 300 BC. His method was sound in principle but weak in practice, and he ended up with a distance equivalent to about five million miles (in truth it is close to 93 million miles). Nevertheless, this became the value accepted by scholars for the next two thousand years. Not until the early days of modern astronomy, in the seventeenth and early eighteenth centuries, did astronomers begin casting around for some better way of finding the sun's distance.

It was the English astronomer Edmund Halley, soon to become Astronomer Royal, who in 1716 first showed explicitly how the transits of Venus could be used to find the distance to the sun. Others before him had vaguely suggested the possibility without examining the details, but it was Halley who supplied not only these, but also lent the whole idea the benefit of his own immense prestige. Halley was almost sixty at the time, and he realized that he himself could not hope to observe the next transit in 1761, but he was successful in kindling a fire of enthusiasm that was later to be transmitted by the French astronomer Joseph-Nicholas Delisle and eventually sweep up most of the astronomers of Europe.

At the time, only one person had ever observed a transit of Venus. Or, more accurately, one person and his friend. This was an extraordinary young Englishman, Jeremiah Horrocks, and his friend William Crabtree, back in the year 1639. I call him extraordinary because, although he was only a very young curate in a very obscure village in Lancashire, and although he was probably nothing at all to his contemporaries, he touched on virtually all of the astronomy of his day in a way which shows astonishing insight. In many ways he was generations ahead of his time; indeed we do not know the full extent of his genius, because he was too poor to publish anything himself, and most of his manuscripts were lost before the remainder were published many decades after his death. Just his very few years of adulthood (he was dead at twenty-three) were sufficient to earn him a place in history, and there is little doubt that had he lived he would have become one of the legendary figures of astronomical history.

In the autumn of 1639 Horrocks discovered from his studies of astronomical tables that Venus should transit the sun sometime in late November of that year. The predicted date varied somewhat depending on whose tables he used, so characteristically he sat down and calculated a more firm date for himself. It turned out to be Sunday, November 24, an unfortunate day of the week for a busy curate. Perhaps with that in mind, Horrocks wrote to his friend William Crabtree, thirty miles away in Manchester, and invited him to make the observation also. Mr. Crabtree was a linen draper, but as Horrocks said, "in mathematical knowledge [he] is inferior to few."

Careful preparations were made, reflecting Horrocks' grasp of what would be important, and finally the day came:

When the time of the observation drew near, I retired to my apartment [wrote Horrocks]; and having closed the windows against the light, I directed my telescope . . . thro' the aperture towards the Sun. . . . I observed the Sun from the time of its rising to 9 o'clock; and again, from a little before ten until noon; and at one in the afternoon, being called in the intervals to business of the highest moment, which for these ornamental pursuits, I could not with decency neglect.

This business of the highest moment was Horrocks' regular Sunday duties; he had to take Matins, Holy Communion, and Evensong, and preach two sermons! One can imagine the sermons being a little briefer than usual, but even so sunset came early at that time of year (the date, incidentally, would have been December 4 by our modern calendar), and just to add to the curate's agitation clouds began to clutter the sky.

> But at 3h 15m in the afternoon, when I was again at liberty to continue my labours, the clouds, as if by *Divine Interposition*, were entirely dispersed, and I was once more invited to the grateful task of repeating my observations. I then beheld a most agreeable sight, a *spot*, which had been the object of my most sanguine wishes . . . just wholly entered upon the Sun's disk. . . . I was immediately sensible that this round spot was the planet *Venus*. . . .

It is a comment on Horrocks' abilities that whereas his predecessors had been wrong by days in their predictions, he himself had not only predicted the event for November 24, but for 3 pm on that date!

Meanwhile, over in Manchester, Mr Crabtree had also been having at least partial success, although he, poor chap, was so overcome by ecstacy at the sight that he failed to make the necessary measurements as instructed by Horrocks:

> Mr Crabtree . . . intended to observe the transit in the same manner with me; but the sky was very unfavourable to him, and was so covered with clouds, almost during the whole day, that he gave himself up entirely to despair and resolved to take no further trouble in the matter. But a little before the time of sun-set, about 3h 35m by the clock, the Sun breaking out for the first time from the clouds, he eagerly betook himself to his observation, and happily saw the most agreeable of all sights, *Venus* just entered upon the *Sun*. He was so ravished with this most pleasing contemplation, that he stood viewing it leisurely, as it were; and from an excess of joy, could scarce prevail upon himself to trust his own senses. For we astronomers have a certain *womanish* disposition, distractedly delighted with light and trifling circumstances, which hardly make the least impression on the rest of mankind. Which levity of disposition, let those deride that will. . . .

For all his talents, the Rev. Horrocks was not above a little male chauvinism.

Of course, these observations were of no help in determining the sun's distance, being not only too crude, but also being made from only one place on the earth. Horrocks, however, did foresee that the transits might one day be used for this purpose.

Incidentally, Horrocks and Crabtree had never met, being too poor to afford the thirty-mile journey between them. They did, however, arrange to meet about a year after the transit, only to have Horrocks die "very suddenly" the day before. And Crabtree himself was soon dead, a victim of the Cromwellian War.

Between the transit of 1639 and that of 1761 science underwent a very profound transformation: it became organized. Until about 1660 science was strictly an individual effort; people like Horrocks and Crabtree and the innumerable others who went before them, worked alone, and only if they had sufficient means as individuals could they afford to publish their results and make them known to other like-minded men. This was the essential tragedy of Jeremiah Horrocks, a genius with no means of communication. But around the middle of the seventeenth century people with an interest in science began to band together to form national societies which met regularly to hear news of scientific happenings, and which published their proceedings and transactions so that others might read of what was happening. Without this immensely important change modern science could not have burgeoned, and the great journeys of the early peripatetic astronomers would never have taken place.

Although such societies were formed in many places in Europe, there were two that were pre-eminent and of particular importance to astronomy: the Royal Society of London and the Académie Royale des Sciences in Paris.

The Royal Society had its origins in the informal meetings of a group of scientifically-minded gentlemen (such as the architect-astronomer Christopher Wren) who gathered as early as 1645 at Gresham College in London to hear of each other's experiments and ideas. The formal constitution of this group as the Royal Society, however, had to await the conclusion of the English Civil War and the restoration of the monarchy, so that it was not until November of 1660 that the Society received its royal charter. Its function, noted its first secretary, was "To improve the knowledge of natural things, and all useful Arts, Manufactures, Mechanick practices, Engynes and Inventions by Experiments (not meddling with Divinity, Metaphysics, Moralls, Politicks, Grammar, Rhetorick or Logick)." Early science was rarely free of commercial overtones. The Society received very little in the way of financial support from outside, and the Fellows were forced to pay dues to maintain their Society. Although this had its disadvantages, it gave them a freedom from governmental interference that they would always hold to with a fierce pride.

The French Academy had rather different beginnings. It too had its nucleus of interested individuals, but its foundation was essentially the work of the French statesman Jean-Baptiste Colbert, trusted advisor to Louis XIV. Colbert saw the Academy primarily as a way of increasing the prestige of his great country, and went about staffing it like the modern owner of a professional sports team. Europe was scoured for the top men in every field of science, and invitations went out to such figures as Newton in England, Leibniz in Germany, Huygens in Holland, Römer in Denmark, and Cassini in Italy. Acceptances by many of these when the Academy was founded in 1666 gave it an almost immediate eminence that the Royal Society would take much longer to acquire. Colbert went on to make working conditions as attractive as possible. Not only was each academician paid a salary, but additional funds were provided for their research work. With this, however, went the proviso that all official addresses and the names of newly proposed candidates would have to be previously approved by the French Government.

This organizing of the top scientists in many countries soon brought into focus the most urgent scientific problem of the day, a problem that went far beyond science, in fact, and was very pressing from a purely commercial point of view: the problem of navigating at sea. Determining latitude at sea was a relatively simple matter, but finding a ship's longitude had long seemed impossible. The difficulty was that while latitude can be found from the measurement of angles alone, longitude requires the use of timekeeping, and no clock had yet been invented that would keep time at sea. There was a chance, though, that some method involving the use of a natural timekeeper, such as the movement of the moon among the stars, could be developed in place of a man-made clock. Such an astronomical problem lay clearly within the realm of the new scientific societies, and these, at the urging of their royal patrons and shipping interests alike, soon turned their attention to the matter.

But to find a solution would require instruments to make observations of the various celestial phenomena, and so very soon after the formation of the scientific societies there was a cry for the building of national astronomical observatories.

The French were first off the mark, and only a year after the Academy had been formed the groundwork was laid for a Royal Observatory in Paris. Colbert was determined that it should reflect the magnificence of his king, so Claude Perrault, designer of the Palace of Versailles, was made architect, with instructions to produce a spacious observatory with the best of everything, including sumptuous living quarters for the resident astronomers. He did not fail.

The British, as usual, were far more parsimonious. The Royal Observatory at Greenwich, founded in 1675, was designed by Christopher Wren, but its construction was out of old bricks from a ruined building on the site and wood from a demolished gate-house in the Tower of London. Money to pay the workmen and any other expenses must come, said the Royal Warrant, "out of such monies as shall come to your hands for old and decayed [gun]powder, which hath or shall be sold by our order of the 1st of January last, provided that the whole sum, so to be expended and paid, shall not exceed five hundred pounds." The first Astronomer Royal, appearing in the warrant under the title of "astronomical observator", was paid £100 a year (£10 withheld for taxes), and had to provide his own instruments. He was allowed one semi-skilled assistant.

Neither of these great observatories proved of much use in determining longitude at sea. That had to wait almost a century for the invention of the marine chronometer. What the observatories did accomplish, though, was a great deal of fundamental astronomy that later proved of immense benefit. And, of course, through the Royal Society and the French Academy, they were very much involved in the transits of Venus. Fascinating contrasts (and some hair-raising stories) emerge from the look we shall take at six of the expeditions, three French and three British, sponsored by the two societies to observe the transits of 1761 and 1769.

In the late 1750s the assistant astronomer at the Greenwich Observatory was Charles Mason. His motives for going on a voyage to observe the

1761 transit are not very difficult to guess. For one thing, he would probably have lost his job by refusing the Royal Society's "invitation", but quite apart from that he may well have seen the transit as an opportunity to escape the drudgery of his position.

The lot of the Greenwich assistants was for long a sorry one. The attitudes of their chiefs towards them can be summarized in the words of one Astronomer Royal:

> I want indefatigable, hard-working, and, above all, obedient drudges (for so I must call them, although they are drudges of a superior order), men who will be contented to pass half their day in using their hands and eyes in the mechanical art of observing, and the remainder of it in the dull process of calculation.

And if the job was dull, there was not a great deal to look forward to outside the job. For one of Mason's social level in the mid-eighteenth century, meals would have been based largely on bread and cheese and beer, with boiled beef and potatoes on occasion, while amusements were circumscribed at best, running to cockfighting, gambling, and the excitement of attending public executions on the eight annual hanging days. Nor was Mason well-paid; he may have reflected with some bitterness that his annual salary of £26 was precisely the same as that paid his counterpart of almost a hundred years earlier when the observatory was founded. For Charles Mason, then, thirty-two years old, healthy and unmarried, the thought of leaving Greenwich to travel the world must have caused no small excitement.

Mason's invitation from the Royal Society was only one of many. For this pair of transits (1761 and 1769) the British would send out observers to sites as widely separated as northern Canada, the south Pacific and South Africa, the French to Siberia, the Indian Ocean and Mexico, and other countries to similarly separated sites. All told, more than a hundred observers were involved.

The hazards they would face were enormous. Sea travel in the little ships of the times was usually a grim business, navigation was still far from perfected, and great gaps on the charts testified to major portions of the world still unexplored. Not even the outlines of Australia were fully known. To add to these hazards, many European countries, including the two superpowers of the day, Britain and France, were at war with one another in the early 1760s, so that there was also the worry of being captured or sunk by an enemy fleet. It is a remarkable testimony to the observers' courage that they rarely failed in their tasks.

Charles Mason was picked to be the assistant astronomer on an expedition to the island of St Helena in the South Atlantic headed by Nevil Maskelyne, a prominent British astronomer who would five years later become Astronomer Royal. Their preparations were already well under way when there occurred an event that would eventually transform Mason's name from total obscurity to a household word in the centuries to come.

This event was a decision by the Royal Society at almost the last minute to send out yet another expedition, to Bencoolen in Sumatra. There was a hasty reorganization of personnel, and Mason found himself with the offer of leading the expedition to the Far East. As leader he would receive an increase in both pay and liquor allowance. He accepted. The remaining problem was to find him an assistant. A beating of the astronomical bushes finally produced, not a fully qualified astronomer, but a land surveyor from County Durham in Ireland, whose life until then had been a good deal less fascinating than the fact of his birth down a coal mine. But he too would find fame, for his name was Jeremiah Dixon.

Having to travel so far, Mason and Dixon left England in December 1760, six months before the transit was due. They could hardly have expected to set foot in England again for another year, but in fact it was only a matter of hours before they were back in port amid 11 dead and 37 wounded, the result of a short but sharp set-to between their ship, *The Seahorse*, and a French frigate *le Grand*.

The British Admiralty immediately set to work on repairing *The Seahorse*, and proposed next time sending it out with a quite considerable escort. This enthusiasm, however, was not matched by any on the part of Mason and Dixon. Mason got off a prompt letter to the Royal Society saying that the delay occasioned by the repairs would undoubtedly make it impracticable to get to Bencoolen in time, and so how about a site much nearer home? He was frank enough to say that not only was there that aspect, as well as the French navy, but a more common maritime problem as well: ". . . the Uneasiness I have my self, to see the Uncommon Misfortune that have attended our designs, and the sea sickness besides; have affected me in an Unusual Manner." Just to be quite clear on the point, Mason's letter said flatly, "We will not proceed thither, let the Consequence be what it will."

There was now an exchange with the Royal Society that was as short and as sharp as that they had had with the French frigate, and once again Mason and Dixon lost. The Royal Society's view came back by return mail, indicating great surprise at the stand they had taken, and noting that ". . . their refusal to proceed upon this Voyage, after their having so publickly and notoriously ingaged in it . . . [would] be a reproach to the Nation in general, to the Royal Society in particular, and more Especially and fatally to themselves. . . ." Additionally, it could not ". . . fail to bring an indelible Scandal upon their Character, and probably end in their utter Ruin." And just to be quite clear in its turn, the Society noted that it would, "with the most inflexible Resentment" prosecute Mason and Dixon in court "with the utmost Severity of the Law" if they failed to set out for Sumatra. Mason and Dixon began to see things in a different light, and on February 3, 1761, a short note to the Royal Society announced that "their dutiful servants" would sail for Sumatra that very evening.

They managed to avoid the French this time, but it still took almost three months just to get as far as the southern tip of Africa. There Mason got off another letter to the Royal Society triumphantly announcing that Bencoolen had been taken by the French and that he and Dixon would be staying right where they were to observe the transit. And, as it happened, just as well too. Whatever might have befallen them in Bencoolen, conditions at Cape Town on the day of the transit were absolutely perfect, and Mason and Dixon obtained a very valuable observation. It was the only one from the South Atlantic region, the St Helena expedition, for instance, being clouded out.

Unlike their modern counterparts, the travelling astronomers of those days were not ones to dash off home the day after the event. Mason and Dixon stayed on in Cape Town for some time, carrying out a measurement of the earth's gravity there, and determining the latitude and longitude of that city with such precision that for long thereafter its position relative to Greenwich was much more accurately known than were the positions of many European cities.

It may well have been the reputation they so gained for excellent work that led to their being called to the American colonies in 1763 to survey the boundary that became famous as the "Mason-Dixon Line."

Six years later a second transit of Venus claimed their attention, but what with the sea-sickness and all, neither was too anxious to make any lengthy voyages. The team split up, Mason going to Ireland for the event, and Dixon to northern Europe. Voyages never were much good for Mason; in 1786 he decided to return to America, was severely stricken at sea, and died in his fifty-sixth year.

British America had its transit observers too. The most important of them was probably William Wales, who had previously been employed by the Royal Greenwich Observatory as a computing assistant. Having heard of some of the frightful happenings that had befallen a number of observers in 1761 (as we shall see), he announced his willingness to be an observer in 1769 provided he was sent to a warm and not too out-of-the-way place. The Royal Society picked him for Hudson Bay in northern Canada. He and his assistant, Joseph Dymond, would be paid £200 *after* they had returned from their year-long stay there.

Wales is described by one of his contemporaries as "a good man of plain, simple manners, with a large person and a benign countenance," with the added note that "there was in William Wales a perpetual fund of humour, a constant glee about him. . . ." Qualities that were to be sorely tried in the coming months.

Because of the very short shipping season in Hudson Bay, it was necessary for Wales and Dymond to leave England in the summer of 1768 and to winter over at their chosen site, Fort Churchill, before the transit in June of 1769. Just getting there was interesting enough, as we may judge from Wales' journal. They had sailed from England at the end of May 1768, and a month and a half later were approaching the Hudson Straits off the northern tip of Labrador. . . .

*July 16.* The former part of these 24 hours we ran through several very strong riplings of the tide, which made us suspect that we might be nearer the entrance of the Straights than our accounts shewed us to be; and therefore about 11h the whole fleet brought to, as the fog was exceeding thick. . . .

*July 18.* This day, and yesterday, we have run through several very strong riplings of tide; and have passed by many islands of ice; but their distance, and the thickness of the fog, rendered it impossible for me to give any account of them.

*July 25*. This day, as I was observing the sun's meridional altitude, there came alongside of us three Eskimaux in their canoes, or, as they term them, Kiacks. . . .

The men have on their legs a pair of boots, made of seal skin, and soled with that of a sea horse; and above these they have breeches made of seal or deer skin. . . . The remaining part of their cloathing is all in one piece, much in the form of an English shift. . . .

The dress of the women differs not from that of the men, excepting that they have long tails to their waistcoats behind, which reach quite down to their heels; and their boots come up quite to their hips, which are there very wide, and made to stand off from their hips with a strong bow of whalebone, for the convenience of putting their children in. I saw one woman with a child in each boot top.

*July 27*. This evening I told 58 islands of ice, all going directly across the Straits . . . at the rate of several miles per hour. . . .

*July 29*. At 15h we hauled the wind to the southward, the ice being quite thick a-head of us. It consisted of large pieces close jambed together; in the place where we attempted to pass through, it was not quite so close. It is really very curious to see a ship working amongst ice. Every man on board has his place assigned him; and the captain takes his in the most convenient one for seeing when the ship approaches very near the piece of ice which is directly a-head of her, which he has no sooner announced, but the ship is moving in a quite contrary direction to what it was before, whereby it avoids striking the piece of ice, or at least, striking of it with that force which it would otherwise have done. In this manner they turned the ship several times in a minute; the wind blowing a strong gale all the time. . . .

*July 30*. This evening I staid upon deck till after midnight, in hopes to have observed the moon's distance from a star; but, after trying for near an hour, I was obliged to give it up, on account of the twilights, which are amazingly bright in these high latitudes. There is another great inconvenience which attends observations of this kind here, viz. a red haziness round the horizon, to a considerable height, rendering the stars very dim; but at the same time large, something like the nucleus of a comet.

*Aug. 7*. About 5 saw the low land of Cape Churchill, bearing from the S. to S.W.b.S. but the haziness of the horizon made the land put on a different appearance every 4 or 5 minutes. I cannot help taking notice of one circumstance, as it appears to me a very remarkable one. Though we saw the land extreamly plain from off the quarter

"A North-west View of Prince of Wales' Fort in Hudson's Bay"—sketch by Samuel Hearne, 1777 (Public Archives of Canada)

deck, and, as it were, lifted up in the haze, in the same manner as the
ice had always done; yet the man at the mast head declared he could
see nothing of it. This appeared so extraordinary to me, that I went
to the main-top-mast-head myself to be satisfied of the truth thereof; and though I could see it very plain both before I went up, and
after I came down, yet I could see nothing like the appearance of
land when I was there.

They finally landed on August 10 and met the governor, a Mr Moses
Norton, who "behaved to us with great civility, and kindness." And after
breakfast the surgeon of the factory took them out for a walk to survey
the desolation of the surrounding marshlands. They hoped to find a piece
of land suitable for growing vegetables, but "in all that extent we did not
find one acre, which, in my opinion, was likely to do it."
What they did find was this:

I found here three very troublesome insects. The first is the mos-
chetto, too common in all parts of America, and too well known, to
need describing here. The second is a very small flie, called (I suppose on account of its smallness) the sand-flie. These in a hot calm
day are intolerably troublesome: there are continually millions of
them about one's face and eyes, so that it is impossible either to
speak, breathe, or look, without having one's mouth, nose, or eyes
full of them. The third insect is much like the large flesh-flie in England; but, at least three times as large: these, from what part ever
they fix their teeth, are sure to carry a piece away with them, an
instance of which I have frequently seen and experienced.

During the rest of August Wales and Dymond were kept busy erecting
their observatory. This they had brought with them in bits and pieces,
someone in England having fortunately recalled in time that there was no
suitable timber in the vicinity of Fort Churchill. There were occasional
diversions, such as on "the 22d and 23d, the people were allowed to
write to their friends in England, so I employed myself to the same
purpose."
On September 8 an ominous note appears in the journal: "This morning the snow was about two inches deep on the plains." It was time to
prepare for winter. Wales now gave up keeping a diary on a day-to-day
basis, but subsequently wrote up a lengthy account of their life through
the long boreal winter, and we may let him speak for himself:

Much about this time, likewise, we who stayed at the factory began to put on our winter rigging; the principle part of which was our toggy, made of beaver skins: in making of which, the person's shape, who is to wear it, is no farther consulted, than that it may be made wide enough, and so long that it may reach nearly to his feet.

November the 6th, the river, which is very rapid, and about a mile over at its mouth, was frozen fast over from side to side, so that the people walked across it to their tents; also the same morning, a half-pint glass of British brandy was frozen solid in the observatory. Not a bird of any kind was now to be seen. . . . We now killed two or three hogs which captain Richards had been so kind to leave with the governor, which before they well opened, and cut into joints, were froze like a piece of ice, so that we had nothing to do but to hang them up in a place where they would remain in that state, and use them when we thought proper. We used some of these, I believe, in the month of May, which were as sweet as they were the moment they were killed. . . .

In the month of January, 1769, the cold began to be extremely intense: even in our little cabbin, which was scarcely three yards square, and in which we constantly kept a very large fire; it had such an effect, that the little alarm clock would not go without an additional weight, and often not with that. The head of my bed-place, for want of knowing better, went against one of the outside walls of the house; and notwithstanding they were of stone, near three feet thick, and lined with inch boards, supported at least three inches from the walls, my bedding was frozen to the boards every morning; and before the end of February, these boards were covered with ice almost half as thick as themselves. Towards the latter end of January, when the cold was so very intense, I carried a half-pint of brandy, perfectly fluid, into the open air, and in less than two minutes it was as thick as treacle; in about five, it had a very strong ice on the top; and I verily believe that in an hour's time it would have been nearly solid.

It was now almost impossible to sleep an hour together, more especially on very cold nights, without being awakened by the cracking of the beams in the house, which were rent by the prodigious expansive power of the frost. It was very easy to mistake them for the guns on the top of the house, which are three pounders. But those are nothing to what we frequently hear from the rocks up the country, and along the coast; these often bursting with a report equal to that of many heavy artillery fired together, and the splinters are thrown to an amazing distance.

On Sunday, March 19th, it thawed in the sun, for the first time. . . .

The latter end of April, the hunters began . . . to prepare for the spring goose season, which is always expected to begin about that

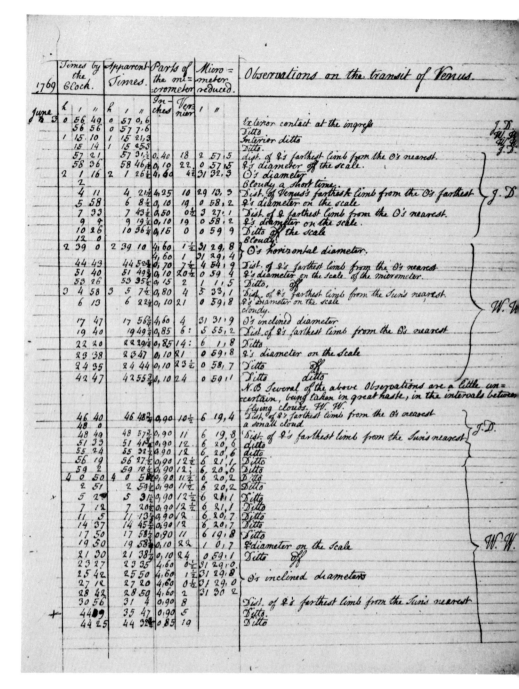

William Wales' handwritten observations on the 1769 Transit of Venus, together with his comments, taken from his "Astronomical Observations By Order of the Royal Society . . ." (reproduction from the original MS by courtesy of the David Dunlap Observatory, University of Toronto)

# Remarks

1. All the measurements of Venus's diameter: and also all those of the Sun, which are not said to be horizontal, were taken with the micrometer, in the same direction that the last preceding distance of the limbs of Venus & the Sun was measured with.

2. We were obliged to alter the rack-work of the micrometer before we began to measure any distances of the limbs &c. in order to make it take in the diameter of Venus, off the scale.

3. The Heavens at the beginning, and for a considerable time both before & after, were frequently obscur'd by clouds: but in the intervals, the air was very clear, & the Sun's limbs extreamly well defin'd.

4. Soon after Venus was half immerg'd, a bright crescent, or rim of light, encompassed all that part of her circumference which was off the Sun; thereby rendering her whole periphery visible. This continued very bright until within a few minutes of the internal contact, & then vanish'd away gradually.

5. We took for the instant of the first internal contact, the time when the least visible thread of light appear'd behind the subsequent limb of Venus: but before that time, Venus's limb seem'd within that of the Sun, & his limb appear'd behind hers in two very obtuse points, seeming as if they would run together in a broad stream, like two drops of oil; but which nevertheless did not happen, but joind in a very fine thread, at some distance from the exterior limb of Venus. This appearance was much more considerable at the egress than at the ingress; owing, as we apprehend, to the bad state of air at that time. We took for the instant of internal contact, at the egress, the Time when the thread of light disappear'd before the preceding limb of the planet, from which time W. W. took notice that he had told about 24" when the limbs of the Sun & Venus were
were apparently in contact; a circumstance which he did not venture to attend to at the ingress.

6. We saw nothing like the appearance of an atmosphere round Venus (unless the above mention'd phænomena may be thought to proceed from thence) either at the beginning, end, or during the time of the transit, nor could we see any thing of a satellite; though we looked for it several times.

7. It may not be improper to add, that the Haziness, complain'd of at the egress, was not owing to any accidental bad quality of the air at that time: it is continually so here to 10° or 12° above the horizon, of & often even to 16° or 18°, in what may be called the clearest state of the heavens.

time; and is, in truth, the harvest to this part of the world.

Toward the latter end of May, the country began to be really agreeable; . . . and the dandelion, having grown pretty luxuriant, made most excellent salad to our roast geese.

On June 16th, the ice of the river broke up, and went to sea; we now set our nets, and caught great plenty of fine salmon; I have known upwards of 90 catched in one tide. . . .

But the return of summer brought with it the "muschettos," which were troublesome to such a degree that Wales, frozen brandy and little sleep notwithstanding, was moved to write, "I cannot help thinking that the winter is the more agreeable part of the year. . . ."

Happily, however, after all that, the day of the transit was clear, and Wales and Dymond made their observations without any hitch. It would still be another three months before they could sail again for England, but when they did, as a final flourish to their success, they were rewarded with the spectacle of a fine comet, which Wales observed throughout most of the voyage home.

Their arrival in England, in mid-October 1769, was marred by the British bureaucracy. Wales was particularly proud of the heavy seal-skin winter clothing that he had acquired through the Hudson Bay Company in Churchill, and no doubt looked forward to its use through winter nights of observing in the future. Instead, every single article, except for some reason his snow-shoes, was confiscated by a customs officer, and despite Wales' subsequent outraged cries to the authorities, none of it was ever returned. Wales was later to sail the world as an astronomer with Captain Cook; one hopes he learnt to deal more effectively with customs officials.

While Wales and Dymond were huddled over the fire in their little "cabbin," others were having a much better time of it elsewhere. In the South Pacific for instance.

Some years before the 1769 transit the Royal Society began to think of sending astronomers to some point in the South Pacific for the event. Not only was that region particularly desirable for the purpose of the transit, but an expedition there would have ramifications extending far beyond the transit. The general scientific and geographic fruits to be harvested, to say nothing of possible political fruits, would be great indeed.

It was, however, an immense undertaking, possibly involving a circumnavigation of the globe, and at the very least a lengthy voyage through a largely unexplored ocean. Costs, they estimated, would run to a good £4000, quite apart from a ship and crew. All of which was beyond the resources of the Royal Society. But what they lacked in funds they made up for in resourcefulness, and soon a carefully worded document was on its way to their patron, King George III, to see what could be done about the £4000.

The King was known to be particularly patriotic, as well as interested in extending the boundaries of empire through maritime exploration. The latter point was covered by carefully noting that the expedition was to improve astronomical knowledge, on which navigation depended. The patriotic angle entered with a note that "several of the Great Powers in Europe, particularly the French, Spaniards, Danes and Swedes are making the proper dispositions for the Observation thereof: and the Empress of Russia has given directions for having the same observed in many different places of her extensive Dominions. . . . That the British Nation have been justly celebrated in the learned world, for their knowledge of Astronomy, in which they are inferior to no Nation upon Earth, Ancient or Modern; and it would cast dishonour upon them should they neglect to have correct observations made of this important phenomenon. . . .

"The Memorialists, attentive to the true end for which they were founded by Your Majesty's Royal Predecessor, The Improvement of Natural Knowledge, conceived it to be their duty to lay their sentiments before Your Majesty with all humility, and submit the same to Your Majesty's Royal Consideration."

It didn't take His Majesty long to conceive where *his* duty lay, and soon the Society had the needed £4000.

They had just as much luck in the problem of a ship: the Admiralty willingly agreed to the loan of one of theirs. Next was the question of who would command the ship.

The Society thought they had just the man. His name was Alexander Dalrymple, and, although his astronomical attainments were somewhat modest, he had the inestimable advantage of having previously been a professional sailor with the East India Company. In addition Dalrymple was eager to go, seeing himself as some kind of latter-day Christopher Columbus out to discover the great *Terra Incognita*. In fact he was quite firm as to his own role in the event: "However it may be necessary to observe that I can have no thought of undertaking the Voyage as a Passenger going out to make the Observations, or on any other footing than that of having the management of the Ship intended for the Service."

The Admiralty were aghast at the mere suggestion of a civilian commanding one of His Majesty's ships, and indeed the First Lord, Sir Edward Hawke, rather extravagantly announced that he would suffer his right hand to be cut off rather than sign such another commission as had gone to Edmund Halley back in 1698. (Halley had wangled command of a Navy ship, the *Paramour Pink*, to undertake a scientific voyage around the Atlantic, and at Barbados had met with a mutiny on the part of his first lieutenant, an experienced naval officer. The subsequent Court of Inquiry, while nominally finding the mate at fault, clearly took a dim view of Halley's role.)

But anyway, if Dalrymple was "totally repugnant to the rules of the Navy," someone else would have to be found. After some wrangling, the Admiralty offered the services of one of their more junior officers, a man who was making a reputation for himself as a marine surveyor and navigator on the St. Lawrence River in Quebec and around the coasts of Newfoundland. His name was James Cook, the ship he was to command the *Endeavour*. And so, through the transits of Venus, began Cook's great voyages of exploration through the Pacific.

On August 26, 1768, at two o'clock in the afternoon, the little 370-ton *Endeavour* slipped out of Plymouth harbour and started south. On board was Mr Charles Green, assistant astronomer at Greenwich, who was replacing the "repugnant" Dalrymple, and also a remarkable young man named Joseph Banks.

Banks was one of those larger-than-life Englishmen who in later years as Sir Joseph and President of the Royal Society would come to dominate the scientific life of late eighteenth-century England. A young gentleman of immense wealth and leisure, he had gone up to Oxford to study botany. There he found a Professor of Botany who gave no lectures, did no tutoring, offered, in fact, no guidance in the subject at all, and who found it quite disagreeable that there should appear on his doorstep an undergraduate who actually wished to study botany. Characteristically Banks solved the problem by obtaining permission to import, at his own expense, a botanist from Cambridge to teach him.

He knew little and cared less for astronomy, but saw the voyage as a golden opportunity for studying the flora of an unknown part of the world. Indeed, the flora of the South Pacific became something of a life's mission with Banks. He made more than one voyage there with Cook, on one of which he encountered one of Cook's ship-masters, a certain William Bligh. It was Banks who in later years sent Captain Bligh off aboard the *Bounty* to bring breadfruit from the Pacific islands to those of the Caribbean, the voyage that was to end in the most famous mutiny of all.

Meanwhile, here he was on the *Endeavour,* nominally off to observe the transit of Venus. He was to become something of a *bête noire* to Cook, forever holding things up and wanting a change of course to gather some new botanical specimen; there was even an occasion when the normally even-tempered Cook exploded, "Damn and blast all scientists!"

In the initial plans for the expedition it was not at all clear just where they would end up observing the transit. There was talk of heading for a group of islands known as the Marquesas, but since the position of these had been lost they would first have to be rediscovered. However, while the *Endeavour* was fitting out there arrived back in England the *Dolphin,* captained by Samuel Wallis, from a circumnavigation of the world. The prime discovery they had to report was of a fabulous island in the South Pacific, an island "such as dreams and enchantments are made of, real land though it was: an island of long beaches and lofty mountains, romantic in the pure ocean air, of noble trees and deep valleys, of bright falling waters. Man in his cool dwellings there was not vile. . . ." It was Tahiti. The major hazard, from a captain's viewpoint, appeared to be the women. These, the word soon spread, had virtually no sexual inhibitions at all, the only price of love being anything metal. The *Dolphin* had almost literally fallen apart where she lay as the eager sailors enthusiastically ripped the nails out of her.

The *Endeavour* would go to Tahiti.

Down the Atlantic they sailed, across to Rio, where they met with a chilly reception from the local governor. He could not believe the *Endeavour,* which had been a North Sea colliery ship before the Admiralty bought her, could possibly belong to the British Navy, and he darkly suspected Cook and company to be smugglers or the like. No one would be allowed ashore unless escorted. Cook preferred to remain huffily in his cabin, but the irrepressible Banks clandestinely slipped ashore at night through an open stern window to bring back botanical specimens.

After that came the nightmare of rounding Cape Horn, and then the vast haul across the immensity of the Pacific. Finally, almost eight months after leaving home, the 7,000-foot peak of Tahiti loomed before them, and on April 13, 1769, the *Endeavour* dropped anchor in Matavai Bay alongside what is even today known as Point Venus.

The welcome by the Tahitians was friendly, but Cook soon had his men at work building a fort—Fort Venus—on the Point. Possibly he hoped that hard physical labour would dampen the ardour of the sailors; if so, he was mistaken. Meanwhile, the scientific party moved ashore to take up residence in tents, where the friendliness of the natives was further tested. As Banks reported to his journal: "Soon after my arrival at the tent 3 hansome girls came off in a canoe to see us, . . . they chatted with us very freely and with very little perswasion agreed to send away their carriage and sleep in [the] tent, a proof of confidence which I have not before met with upon so short an acquaintance."

With the fort completed, the astronomical party began to erect an observatory within it, and the precious instruments were brought ashore. The very next day, despite the walls and sentries, one of the important instruments, an astronomical quadrant, was stolen. Charles Green, the chief astronomer, was beside himself, and, pistol in hand, set off down the beach to find the thief. Banks, of course, was at his side, and they spent the entire day tracking down pieces of the quadrant, which the thief had promptly dismantled. Returning late in the day they were met by Cook and a party of marines coming to look for them, and found that in their absence the wrong man had been captured and imprisoned, a chief at that. It required some nimble work on the part of Cook to smooth things over.

As the day of the transit, Saturday, June 3, 1769, began to approach, the astronomers became anxious watchers of the weather. It was none too promising, and the Friday was downright cloudy, yet there was nothing to do but continue with the preparations. Robert Molyneux, one of the observers, wrote in his journal: "The Captain and Mr Green is entirely employ'd getting every thing compleatly ready. I was order'd to prepare for Observation & had a Telescope ready accordingly, every thing very quiet & all Hands anxious for Tomorrow."

Banks' journal tells of the day itself:

Various were the Changes observd in the weather during the course of last night, some one or other of us was up every half hour who constantly informd the rest that it was either clear or Hazey, at day break we rose and soon after had the satisfaction of seeing the sun rise as clear and bright as we could wish him. I then wishd success to the observers Msrs Gore and Monkhouse and repaird to the Island, where I could do the double service of examining the natural produce and buying provisions for my companions who were engagd in so usefull a work.

    . . . After the first Internal contact was over I went to my Companions at the observatory carrying with me Tarroa, Nuna [local chiefs] and some of their cheif atendants; to them we shewd the planet upon the sun and made them understand that we came on purpose to see it.

    . . . before night [we] arrivd at the tents, where we had the great satisfaction that the observations there had been attended with as much success as Mr Green and the Captn could wish, the day having been perfectly clear not so much as a cloud interveining. We also heard the melancholy news that a large part of our stock of Nails had been purloind by some of the ships company during the time of the Observation, when every body was ashore who had any degree of command. . . . This loss is of a very serious nature as these nails if circulated by the people among the Indians will much lessen the value of Iron. . . .

Evidently the sailors had had a Venus of sorts in mind too, despite the day's temperature in the sun standing at 119°.

It is worth quoting the report of Cook himself, because in it he clearly describes a problem that was disturbing transit observers all over the world:

This day prov'd as favourable to our purpose as we could wish, not a Clowd was to be seen the whole day and the Air was perfectly clear, so that we had every advantage we could desire in Observing the whole of the passage of the Planet Venus over the Suns disk: we very distinctly saw an Atmosphere or dusky shade round the body of the Planet which very much disturbed the times of the Contacts. . . . We differ'd from one another in observeing the times of the Contacts much more than could be expected.

More about this later; meanwhile the intrepid observers had yet to make their way home. They had been gone now for almost a year, and what they would have said could they have foreseen the trials and tribulations they would have to endure for another two years before returning home is beyond imagination.

The expedition remained at Tahiti for more than a month after the transit, with Cook surveying the island and the indefatigable Banks working away at his botany. The astronomers kept themselves amused with such diversions as a banquet to celebrate King George's birthday, and an archery contest with the Tahitians, which soon came to an end when it was discovered that the latter shot for length only, not accuracy.

Eventually, on July 13, 1769, the *Endeavour* hoisted anchor and sailed from the Society Islands, as Cook had named the group in honour of the Royal Society. Cook's task now was not merely to return home; there remained important geographical work to be done, particularly the question of the supposed great southern continent. So they beat away to the southwest and were long occupied with the first complete survey of New Zealand. From there it was westwards again to discover Botany Bay and the east coast of Australia, and sailing north along that coastline they ran into the Great Barrier Reef. Ran into, literally, for their ship became jammed on the razor-sharp coral, and only by the most consummate seamanship was Cook able to prevent the *Endeavour* from foundering completely.

The repairs necessitated by that episode forced them to spend three months in Java, where malaria and dysentery ("the bloody flux") very nearly decimated the crew. When they sailed again, still with twelve thousand miles before them, they were, as one author has said, more like a hospital ship than anything else. At times there were only twelve able-bodied men on board. Away across the Indian Ocean they staggered, around the Cape of Good Hope, and north again up the Atlantic; Banks, though pale and shaky from fever, studying marine life all the way.

At last, on Saturday, July 13, 1771, with hardly half the crew remaining, the *Endeavour* rattled out her anchor in the Downs off southern England. They were home.

As often happens, the survivors must have been surprised to read of themselves in the press. Cook was almost ignored; the newspapers managed to misspell his name, and all the great geographic finds were attributed to Banks. Indeed, Banks and the other "ingenious gentlemen" who had sailed "to discover the transit of Venus" came out of it very well; in between supervising the taking ashore of his 17,000 plant specimens ("of a kind never before seen in this kingdom") Banks was invited to an audience with the King, and the whole thing was the making of his illustrious career.

As for Charles Green, he was one of several astronomers who gave their lives to the transits of Venus. The *General Evening Post* of July 27 reported that he "had been ill some time and was directed by the surgeon to keep himself warm, but in a fit of phrensy got up in the night and put his legs out of the portholes, which was the occasion of his death." Presumably the sharks got him.

It is quite astonishing in reflecting on these and other of the expeditions that went great distances to observe the transits, that despite the enormous hardships and travail of the journeys, and at times the death of the principal observer, the results always managed to reach their destinations back home. It says much for the determination of the people involved. They were making an important contribution to their science, they knew that no one then living would ever again see a transit of Venus; the results *must* get back at all costs.

Nowhere was this better demonstrated than by several of the French expeditions which set out to stalk both the 1761 and 1769 transits.

When the Academy began their detailed preparations for the transits in 1760 Jean-Baptiste Chappe d'Auteroche was a rather fleshy man of thirty-two. He had trained as an astronomer, and done some good work with Cassini de Thury (one of a famous family that directed the Paris Observatory for more than a century), but had led a quiet life that had brought him little prominence. He was eager to travel, so when the Russian Imperial Academy of Sciences invited the French Academy to send astronomers to observe the 1761 transit from Russia, Chappe soon volunteered his services.

The Academy offered him the site of Tobolsk in Siberia, which, perhaps with the innocence of the untravelled, he readily accepted. Initially it was decided that he should go by ship around the top of Europe, and then make his way south through Siberia to Tobolsk. The war with Britain, however, made this a risky proposition, and perhaps Chappe was not too sorry when a delay in his preparations made him miss the Dutch ship he was to have sailed on. He subsequently congratulated himself on this piece of fortune, for the ship ran aground off the coast of Sweden, but there may well have been occasions later when he wished himself on it.

So now he would go by land. East from Paris to Vienna, north-east through Warsaw to St Petersburg, down to Moscow, and on out into the vastness of Siberia. Considering that it would be a journey of almost four thousand miles, including a crossing of the Ural mountains in the dead of winter, he accepted it with remarkable sang-froid.

What it was like to travel in those days, even within civilized Europe, may be judged from the fact that it took over a week just to reach the border of France at Strasbourg, by which time his carriages (there was an enormous amount of equipment with him) were so badly battered as to need replacing, and all his thermometers and barometers were broken. While fresh vehicles were summoned, Chappe had to build new instruments for himself, and in the course of doing that it occurred to him that river travel would be a lot smoother.

Over to Ulm he went, and despite the short daylight hours and heavy fogs (it was December 1760), embarked on the Danube's long, slow, winding course down to Vienna. At times he was held up for days, when he would go scrambling up any nearby mountains with his barometers to measure altitudes, and even on travelling days he occupied himself with mapping the river.

He was in Vienna by the beginning of January, and spent a week there in scientific pursuits with the local astronomers, leaving on January 8, 1761, in temperatures well below zero Fahrenheit, to begin the arduous journey towards Warsaw. The going was again extremely rough, with many accidents and broken instruments, and the crossing of the semi-frozen rivers particularly difficult. Even at fords, Chappe had to go ahead on foot and smash the ice to allow the carriages to cross.

Two weeks later he was in Warsaw, where he heard that the Russian scientists in St Petersburg had practically given up hope of his reaching his post in time. Now he was forced to switch to sleds, and with these made good time towards St Petersburg, reaching there by mid-February.

The Russians had indeed given up hope, and had sent their own observers on ahead. They advised him not to try for Tobolsk, but to settle for some nearer station. Chappe must have been tempted, for he still had an immense distance to go, the journey would probably be rougher than ever with the Urals to cross, and his bulky load would now have to be further increased. From here he would have to carry much of his food

*Jean Chappe d'Auteroche making a halt at a peasant's cottage in the village of Melechina, Siberia, en route for Tobolsk from St. Petersburg. The illustrations for his account of Siberia were engraved by J. B. Le Prince who knew Russia at first hand.*

with him, in fact all his needs down to a bed, and would need an inter-
preter and guides. But Tobolsk was particularly desirable for scientific
reasons, and the hardy Chappe determined to reach it at all costs.

He received excellent co-operation from his Russian colleagues, and
in early March the expedition was off again aboard four giant sleds,
each drawn by five horses running abreast. One can picture them racing
for the Urals through the silence of the frozen Russian countryside, great
clouds of snow rising from the horses' flying hooves. Chappe's greatest
fear was not that the distance would be too much to cover before the
transit, which was still three months off, but that he would be overtaken
by the onrushing spring. To be bogged down in the thaw would mean the
end of his hopes.

On the steaming horses pounded, out past what is now Gorki, where
"the surface of the Volga was as smooth as glass . . . and the sledges
went on with inconceivable swiftness." The only pauses were to change
horses at the infrequent relief posts.

Fortunately, the Urals gave them little trouble, although Chappe
awoke one morning to find that all his guides had deserted him. He was
forced to hunt them down through the Siberian forests and bring them
back at pistol point.

They made it. The final river was crossed only six days before the ice
broke up, and on April 10, 1761, Chappe thundered into Tobolsk.

Wasting no time, he immediately set about erecting an observatory
on a mountain outside the town, and began making the ancillary observations so important to the success of his mission. (He had, for example,
to determine his latitude and longitude with the greatest possible precision.)

Now new trouble arose. The thaw was early that year, and the break
up of ice on the rivers particularly severe. Tobolsk is at the confluence
of two rivers, the Tobol and the Irtysh, and soon the town was being
flooded. Rumours began to spread among the superstitious townsfolk
that this had something to do with the interfering foreigner up on the
hill, who, they knew, had come to fool with the sun in some way. Rum-
blings of mob action arose, and the town's Governor was forced to post
a guard on Chappe and his observatory.

Chappe had been diplomatic enough to make good friends with the
Governor and the local Archbishop as soon as he had arrived, and now,
ostensibly out of thanks for their intervention, but in truth to keep them

out of his way during the actual transit, he set up a special telescope for them to use during the event.

The day approached. Chappe reports how he settled down at the observatory the evening before: "The sky was clear, the sun sunk below the horizon, free from all vapors; the mild glimmering of the twilight, and the perfect stillness of the universe, completed my satisfaction and added to the serenity of my mind." Serenity be damned; the poor fellow was too excited either to eat or sleep that night!

June 6, 1761. For Chappe it dawned clear and bright: "The moment of the observation was now at hand; I was seized with an universal shivering, and was obliged to collect all my thoughts, in order not to miss it. . . . Pleasures of the like nature may sometimes be experienced; but at this instant, I truly enjoyed that of my observation, and was delighted with the hopes of its being still useful to posterity, when I had quitted this life." Words he might have saved another eight years.

Like other transit observers elsewhere, Chappe was in no hurry to get home. He stayed on in Tobolsk another three months, carrying out the usual scientific work ranging from geology to a social study of the people, before setting off on the four thousand miles home. It was a leisurely trip indeed; he spent almost a year in St Petersburg writing up and having printed a memoir on his work, and it wasn't until November of 1762 that he again set foot in Paris.

While in St Petersburg Chappe received an earnest invitation to stay and become a scientist in the Imperial Academy; the Empress even offered him a much-coveted position of seniority there. He declined. Chappe in fact had a very low opinion of the Russian people as a whole. He observed that such fame as had accrued to the Imperial Academy had come entirely from the foreign scientists who had been lured there, and that "not one Russian has appeared . . . whose name deserves to be recorded in the history of the Arts and Sciences." He later published a work elaborating on this, developing a theory of how the Russian climate interacted with the physiology of the people to produce in them coarse minds, although he clearly saw that a lack of education and a despotic government played their roles too. The whole thing so enraged the Empress that she wrote a line-by-line rebuttal of it.

With all that behind him, it is not too surprising that when the 1769 transit began to loom up Chappe rather thoughtfully proposed that this time he go somewhere quite different. Such as the South Seas. That locale was considered by the French Academy, but in the end dropped for political reasons, and instead Chappe found himself on the way to Baja California in Mexico.

Despite the seventy-seven day voyage across the Atlantic to Veracruz, and the lengthy trek right across Mexico to the southern tip of Baja California, the journey itself was relatively uneventful. By mid-May of 1769 Chappe's party was well-established at a Spanish mission in a small village there, and Chappe was busy determining his latitude and longitude. Since the California peninsula contains some of the finest astronomical sites anywhere in the world, there was little doubt that success would attend the venture. And indeed, Chappe's transit observations on June 3 were among the best made.

At this moment of triumph, however, Chappe was abruptly cut down. An epidemic disease swept through the village, killing off three-quarters of the inhabitants, and Chappe himself soon was stricken. He lived long enough to observe a lunar eclipse (important to his longitude determination) on June 18, but within weeks had "quitted this life." Only one of his party survived to take home the precious observations.

When, back in 1760, the French Academy had called for volunteers to go to Siberia, there had been another man besides Chappe eager for the job. This was Alexandre-Gui Pingré. In the end, he and Chappe had amicably agreed that Chappe should go to Siberia, while Pingré would take whatever other voyage might be forthcoming to observe the transit of 1761.

The typical savant of the eighteenth century was, of course, a man of far wider interests than is his counterpart today, but even so, Pingré stands out for the catholicity of his abilities. Born in Paris in 1711, he was trained as a theologian, and at the remarkably young age of twenty-four he became a professor of theology. It seems, however, that the hierarchy of the Church was soon displeased with Pingré's liberal ideas on many topics, and before long he found himself downgraded to being a mere country school-teacher. Having considerable mathematical abilities, he now switched to the study of astronomy, and before long began to make something of a name for himself in this field. Curiously enough, his rising reputation as a scientist served to restore him in the eyes of the Church, and by mid-century he had returned to Paris to become librarian to his Order of Sainte Geneviève. Here he settled down to a double professional life; on the one hand turning out an extraordinary amount of ecclesiastical work, poetry, literary sketches, musical satires, etc., while on the other hand gaining an increasing recognition as a first-rate astronomer in the Academy of Sciences. His literary abilities were such that as late as 1903 the classicist A. E. Housman could say that Pingré's translation of the Roman poet Manilius was still the best available; yet the great bulk of Pingré's literary manuscripts remain unedited and unpublished in the archives of Sainte Geneviève.

Now, at the age of almost fifty, Pingré was eager to travel. At first the Academy thought to send him to somewhere on the coast of South West Africa or Angola to observe the transit. The difficulty of getting him there and back by commercial shipping, however, made them think of some more northerly site on the coast of West Africa. This, again, required the co-operation of the Dutch or Portuguese, which seems not to have been forthcoming, so eventually he was destined for the island of Rodrigue some way east of Madagascar. Not only was this French, but someone recalled that the weather there in June was particularly clear.

He left Paris on November 17, 1760, after an enthusiastic send-off party the night before, although he reported himself to have been too terrified at the forthcoming voyage to have been able to eat very much. He had to go overland to Port Louis on the coast of Brittany to embark, and, as Chappe was finding out, discovered that road travel in France was terribly hard on one's instruments.

Arrived at Port Louis he got into a violent altercation with the shipping agents over the half-ton or so of baggage he proposed taking with him. This was not at all abnormal for an astronomer, he argued hotly, but it took further delays and letters from the Academy to smooth things over.

What he considered to be one of his most valuable possessions at this point, however, hardly added much to his baggage. It was a piece of paper, one of the more remarkable documents of history, and its origins were these. Britain and France were at war with one another at this stage, and the Academy feared particularly for Pingré's safety while out on the high seas. They therefore took what by today's standards would seem the extraordinary step of writing a lengthy appeal to the British Admiralty, explaining that Pingré's voyage was for purely scientific purposes, and asking the Admiralty to issue an order restraining all British ships from molesting Pingré's vessel during its travels. This the Admiralty did, in the form of a letter to be carried by Pingré. It was addressed to The Respective Captains and Commanders of His Majesty's Ships & Vessels & Commanders of Privatiers, By Command of their Lordships, and read as follows:

> Whereas the Academy of Sciences at Paris have appointed two of their members to proceed to different parts of the world, to observe the expected *Transit of Venus over the Sun*; one of whom, the Bearer, *Monsieur Pingré*, is to make such observation in *The Island Rodrigue in the East Indies*; and whereas it is necessary that the said Monsieur Pingré should not meet with any Interruption either in his passage to or from that Island, you are hereby most strictly required and directed *not to molest his person* or Effects upon any Account, but to suffer him to proceed without delay or Interruption in the Execution of his design. Given under our hands the 25th of November 1760.

Viewed from our own age of total warfare, the whole thing has a certain charm to it; can one imagine, for instance, Churchill dropping a line to Hitler asking that the Wehrmacht please not shoot down some plane bent only on scientific purposes?

So, clutching this talisman, Pingré embarked on his voyage south to the Cape of Good Hope and up into the Indian Ocean. It was January 9, 1761, a few weeks after Mason and Dixon had started off on a very similar voyage.

For a while it looked as though the similarities would be closer still, because on their very first day out they sighted a squadron of five British ships. The limitations of Pingré's letter were only too apparent: the British ships would come in, cannons blazing, first, and the reading of any documents would be strictly second. The French captain therefore issued an order to battle stations, which caused much more than the usual stampede. This was because the ship had been modified to accommodate a considerable number of passengers, and the partitions that had been erected for their cabins blocked access to some of the cannons. These were hastily ripped down and all the passengers' belongings flung pell-mell into Pingré's cabin, which somehow escaped the carnage. One can imagine Pingré's expostulations; only the night before he had been complaining about lack of sleep brought on by an attack of gout, and now this.

Fortunately, however, with a shift in the wind and the arrival of darkness they were able to evade the British and proceed on their way.

Pingré spent a good deal of time during the voyage experimenting with various ways of determining longitude at sea. He got into an interesting argument with the ship's navigator, when the latter's calculations showed they would pass the west coast of the Cape Verde Islands, while Pingré claimed they would pass to the east. During this time they hove to at night in case neither was correct and they sail right into the islands. Finally they agreed on a position which, according to their official naval chart, should have placed them right in the middle of the island of Bonavista —nowhere to be seen. Pingré concluded the chart was in error by at least two degrees!

Much of the shipboard time was spent in lesser pursuits. There was an enthusiastic ceremony when they crossed the equator, and the passengers sometimes amused themselves with capturing seabirds and banding them with a variety of witty (not to say racy) epigrams. And, of course, there was always alcohol, about which Pingré was quite unabashed. "Liquor," he firmly announced, "gives us the necessary strength for determining the distance of the Moon from the Sun." Not a very efficient fuel, it seems, since often a good deal of it would be required before the necessary strength was found. Perhaps there were more contributing factors to longitude uncertainties than historians have tended to believe.

By early April the merry spirits had rounded the southern tip of Africa and had started into the Indian Ocean. They were cheered at this point to encounter another French ship, which had recently left the Dutch colony at Cape Town with a fresh load of supplies. Cheer soon turned to gloom, however, when it transpired that this ship, the *Lys*, had skirmished with the British, was damaged, and in need of assistance. The captain of the *Lys* proved to be senior in rank to the captain of Pingré's ship, and demanded that the latter accompany his ship on its slow and laboured return to the islands of Mauritius, then known as Isle de France. Time was running short, and at this pace Pingré would surely not reach Rodrigue in time for the transit, but the new senior officer cared nothing for such frivolities as astronomy.

There now ensued a series of violent Gallic expostulations on the part of Pingré and the two captains, rising in rapid crescendo to an exchange of extremely stiff and formal letters between all parties. Happily, however, an excellent dinner, aided by a liberal flow of the good Cape wines, restored a measure of tranquillity, if not total satisfaction.

They were in Isle de France a month before the transit, and now thankfully changed ships for the remaining few hundred miles to Rodrigue. Flying over Rodrigue en route to Mauritius a few years ago, I noticed that the distance was spanned in about half-an-hour. It took Pingré nineteen days. Becalmed on the way, he must have been beside himself with frustration at the thought of the transit now hardly more than a week away. Finally, with only days to spare, they were on Rodrigue, discovering that the sea air during the long voyage had wreaked havoc on their instruments. It took an almost non-stop effort with the local turtle oil to restore everything to working order, and that left them only time to build the crudest of shelters as an observatory. But they were ready when June 6 arrived.

It rained. Pingré and his assistant must have sat in their leaky little shelter, watching the overcast sky, and even Pingré must have been short on conversation. Nevertheless it did clear partially during the day, and the two astronomers were able to obtain a number of useful observations. Pingré rather optimistically declared the expedition to have been a success, and this called for suitable celebrations at dinner that night. There was a considerable round of toasts to everyone from the King down to just about anyone connected with the transits of Venus.

As was the custom among transit observers, Pingré's expedition remained on Rodrigue for some time doing additional work. Rather longer than they intended, however. Three weeks after the transit a British man o'war sailed in and bombarded and then sacked the island's tiny settlement. They burned one of the two French ships there, and took the other—Pingré's—as a prize.

Pingré, of course, rushed down to confront the British commander with his letter from the Admiralty, only to have the commander contemptuously brush it aside and carry on with the looting. They then departed, leaving Pingré and company stranded on the island until another French ship happened to call three months later. This at least gave Pingré time to compose some extremely vitriolic letters to a host of people, particularly the President of the Royal Society and the British Admiralty, on the subject of their countrymen's gross incivility. Rather interestingly, while the letter-writing was in progress two other British ships visited the island, and while not above doing some more looting, the captains at least had the courtesy to deliver Pingré's mail for him.

It was mid-October before the astronomers embarked from Isle de France on the long voyage home, Pingré praying that tranquillity might attend it. They almost made it this time, but there on February 11, 1762, were the damned British again. Pingré's vessel suffered little damage in the encounter, but it was outgunned and soon captured. They were forced to accompany the British ship into Lisbon, where the weary astronomers were hard put to preserve what little remained of their instruments and natural history collections from the ravages of the British and the port stevedores.

Pingré had had enough of sea travel. He elected to finish the journey overland, even if it meant going by ox-cart across the appalling roads of Spain. It was April 28, 1762, when they at last staggered across the Pyrenees and stared, perhaps with disbelief, at their beloved France once more. They had been gone, Pingré noted heavily, 1 year, 3 months, 18 days, 19 hours, and 53 minutes.

Of all transit observers anywhere, however, the worst victim of sheer, rotten bad luck was Guillaume-Joseph-Hyacinthe-Jean-Baptiste le Gentil de la Galasière.

Born in 1725, Le Gentil had, like Pingré and others, seemed destined for the Church. But while yet a somewhat less-than-ardent theology student, he had gone to hear the great Delisle give a lecture on astronomy, and had been so captured by the subject that very soon he was studying it full-time with Cassini de Thury (as was Chappe). A later Cassini, in writing a Eulogy to Le Gentil many years later, said of him: "M. Le Gentil kept the apparel of abbé only until the title of 'savant' procured him a less equivocal esteem and existence." The Cassinis had a rather dry view of the Church.

Le Gentil's abilities as an astronomer rapidly showed through, and while still in his twenties he was elected to the Academy of Sciences with a first-class reputation already behind him.

Life for the well-to-do in the eighteenth century may have offered many riches, but the opportunity for travel was not one of them. So it was that Le Gentil, like Chappe and like Pingré, saw the forthcoming transits of Venus as an opportunity not only to do something important for his science, but also greatly to expand his experiences. With some eagerness, then, he volunteered his services for the project, and was picked to carry out observations of the 1761 transit at Pondicherry on the east coast of India.

He left on March 26, 1760, heading by the traditional Cape of Good Hope route first for Isle de France and then on to India. His voyage of almost four months to Isle de France has been summed up laconically by one author as being "rather uneventful, save for the loss of a fellow passenger by suicide and the pursuit by an English fleet near the Cape of Good Hope."

It was at this point that good luck began to desert him, for arrived at Isle de France he learnt that Pondicherry was under siege by the British, and that a heavy naval brigade sent from France to lift the siege had been all but destroyed by a sudden hurricane at Isle de France a few months earlier.

Le Gentil settled down to a lengthy stay on the island, at one point toying with the idea of going to Java or Rodrigue instead. But an attack of dysentery dissuaded him from that. The months drifted on, until in March of 1761 a troopship was sent to assist the beleaguered defenders of Pondicherry, and Le Gentil decided to accompany this force.

At first things went well, his ship, *la Sylphide*, making thirty to forty leagues a day running before a favourable wind. But then came the region of the monsoons and they were blown severely off course. Le Gentil explains what followed:

> In this way we wandered around for five weeks in the seas of Africa, along the coast of Ajan, in the Arabian seas. We crossed the archipeligo of Socotra, at the entrance to the gulf of Arabia. We appeared before Mahé, on the coast of Malabar, the 24th of May; we learnt from the ships of this country that this place was in the possession of the English, and that Pondicherry no longer existed for us. Without stopping further, we set sail. I would not yet have despaired if we had followed our first objective to go to the coast of Coromandel; but they made, to my great regret, the resolution to return to the Isle de France.

So, when June 6 arrived Le Gentil was somewhere in the middle of the Indian Ocean. It was a beautifully clear day and he could watch the transit taking place, but there was nothing he could do about it scientifically. For the critical observations were the times when Venus entered and left the sun's disk, and since the only accurate clocks of the day were pendulum ones, there was nothing Le Gentil could do from the heaving deck of a small ship at sea.

The transit may have been a bitter disappointment, but there was certainly no question of rushing straight back home again. Like his colleagues around the world, Le Gentil elected to stay on a while and at least salvage some other scientific work before returning. Based at Isle de France he began a lengthy series of exploratory voyages through the Mascarene islands, up and down the east coast of Madagascar, and elsewhere in the Indian Ocean.

The months slipped into years, and by 1765, far from thoughts of returning home, Le Gentil began to take thought for the forthcoming transit of 1769. Why make the tedious and perilous voyage back to France, only to have to set forth again in a few years? Why not stay out here, where the Academy would undoubtedly want an observer for the 1769 transit?

During his stay at Isle de France, however, Le Gentil had been making additional calculations of the circumstances of this next transit. At Pondicherry, he found, the transit would take place when the sun was very low in the morning sky, whereas a site further east would place the sun higher in the sky and accordingly should yield more accurate observations. For some reason (possibly connected with an opportunity of returning home via Mexico and so completing a circumnavigation of the world) Manila in the Philippines attracted him, and he wrote off to the Academy in Paris suggesting that he go there. While he was waiting out the months for their reply, however, a Spanish warship bound for Manila called at Isle de France, and Le Gentil seized this opportunity for his passage with no more than a hope that the Academy would be agreeable. He had, in any case, had just about enough of life on a small island after six years there:

> I finally left Isle de France May 1, 1766, quite resolved to say goodbye forever to that island; and indeed I had conceived the plan of going back to Europe by way of Acapulco, and thereby finish my trip around the world; but I had not foreseen what was to happen to me at Manila, and that a last adventure had destined me for the Isle de France. We arrived at Manila August 10th. Our voyage was rather long; it had its difficulties and its wearinesses.

Le Gentil liked Manila. It "is without contradiction one of the most beautiful countries in the seas of Asia; the climate is excellent there, the soil is of the greatest fertility. The Philippines have fifteen or sixteen fine ports, and they are covered with the finest woods for building." The people, however—or more specifically the Governor—were something else.

Le Gentil had asked the Academy to send him letters of introduction and recommendation suitable for presenting to the Spanish Governor, and these had now arrived. Indeed, not only was there one from the Academy, but also one from the Spanish Court itself. Le Gentil confidently presented these and received in return a rude shock. The Governor, "a restless man, of evil intentions to the French in general, regarded me . . . with a jealous eye. . . ." He proceeded to do some rapid—and faulty—arithmetic, concluding that there had not been time for such letters to reach Le Gentil, and that they were therefore forgeries.

> This odious and injurious suspicion caused me a good deal of sorrow, and gave me also some worry for the rest of the time which I had still to remain in Manila. . . . I considered . . . that I was running very great risks by staying at Manila. . . . I saw that the governor acted despotically and tyrannically in everything; I saw that, in this distant country, reasons would not be lacking to arrest a man in the course of the most serious and important affairs; I had very striking examples of this before my eyes. . . .

Just to add to these misgivings, another letter from the Academy announced that they were not too happy with Le Gentil's choice of site for scientific reasons, and would prefer that he observe the transit from Pondicherry after all. It was Pingré, now comfortably ensconced in Paris once more, that Le Gentil had to thank for this. Anyway, what with these grumblings and the attitude of the wretched Governor, there seemed to be nothing for it but to set out once again on the long haul for India. Perhaps he later took some comfort if he heard the news that the corrupt Governor was arrested and saw out his remaining days in a dungeon.

This time Le Gentil would travel by a Portuguese ship bound from Macao to Madras. There was a false start when all the cargo slipped and the ship had to return to port for re-ballasting, but since it was only February of 1768 there was no great rush. Even so, as always, hardly anything could be counted on to go right. For example, a rather unpleasant situation developed while they were out in the South China Sea:

We arrived at la Viole at seven-thirty in the evening; it was night, the weather not very clear, and the moon in its course did not lend us its light: we were between two equally redoubtable dangers: one is la Viole, the other a sand bank . . . which is not farther than a league and a half from la Viole. It was necessary to pass between the two, and it was night. I shall confess here to you that I was a little worried. . . .

To recount the affair to you exactly, you will suppose in the first place that in the ships of Macao the captains understand navigation not at all or very little; the first pilot alone is in charge of guiding the vessel, the captain must not meddle in it in any manner. . . . Unhappily for us our captain and our first pilot did not agree very well. . . . We were scarcely through la Viole when the captain cried from above the poop where he was, to the first pilot to make a manoeuver (I do not remember what) of which the latter did not approve; the pilot answered rather brusquely that he would not do it, alleging that he was in charge of the conduct of the ship, that he knew what he had to do, and how it was necessary to navigate. The captain insisted, wanting, he said, to be obeyed. The pilot kept answering him in the same tone: finally the dispute became heated; the latter went into a passion and went and locked himself in his cabin, abandoning his ship to the pleasure of the wind.

Le Gentil, who didn't understand Portuguese too well, at first wasn't sure just what was happening, but the ship starting to drift randomly among the shoals soon brought the realization home, and he rushed down to the pilot's cabin. "He was in it but he was sulking, and nothing that I could say to him was capable of making him take up the helm again."

But if appeals to better nature failed, perhaps economics would not. Le Gentil remembered his fellow passengers, who were Armenian, and whose goods made up most of the cargo. They surely would change the pilot's mind. He dashed to find them and explain the situation, and while the Armenians went to deal with the pilot Le Gentil took over the handling of the ship himself.

The Armenians had frightful difficulty in getting the pilot out of his cabin; but when Seigneur Melchisedek, with that great phlegmatic air . . . had spoken to him about conscience, he finally succeeded in conquering his obstinacy. "*Hombre*", he said to him, (from what he told me a moment afterwards), "*tiene usted conscientia?*" "Are you a man who has any conscience?" He submitted at this word conscience, coming from the mouth of an Armenian, and he resumed the conduct of the vessel. This pilot was a brusque and gross man, he became moody very easily; otherwise it appeared to me that he understood this sort of voyage very well.

Later they somehow acquired two pilots, both of whom gave trouble together on an occasion when the ship was near a small uninhabited island. Le Gentil recounts this horror story:

> Our two first pilots and a passenger had gone to land in the ship's boat, incited by curiosity to see the island. They tried to get me to go to land with them; but my policy is never to quit my ship unless it is in port or in a sure roads; although it was very fine weather when they descended, they could not persuade me. . . . How thankful I was for my resistance when I saw the bad weather and when I perceived all the horror of the condition to which the travellers were reduced! They had astonishing trouble in getting back on board ship: . . . it was darkest night, and there was very heavy rain; they were led only by the waves of the sea which appeared all on fire. . . . Their repeated cries in the middle of the night, the bad weather, the noise of the sea, the efforts they were making to come alongside, all that represented for me the picture of shipwrecked people. . . .
>
> You see by this tale what Portuguese vessels are. I have never heard of a European vessel anchored in an open roadstead, in which there was so little discipline that the two first pilots could thus abandon their ship for a pleasure jaunt. Only the captain remained, and he was as little in condition to conduct his vessel as I am to lead an army, and for pilots, two old automatons to whom I would not have entrusted the conduct of a launch.

It says much for the state of sea travel in the 1760s, though, that Le Gentil, once safely arrived in India, summed up this trip by saying "It is not possible to have a more fortunate voyage than that"! Mind you, he was fast becoming something of an expert on the hazards of the sea, although his worst was yet to come.

At Pondicherry, meanwhile, everything was going very well. The war with Britain was long since over, the town was back in French hands, and the local Governor could hardly have been more solicitous or courteous. He met Le Gentil on arrival, ordered that the latter's goods be off-loaded with great care, and then carried him off to the Governor's country mansion "where I found a large and pleasant company, good music, and an excellent dinner."

It was still only late March of 1768, with more than a year to go before the transit, but the very next day the Governor invited Le Gentil to set about choosing a site for his observatory, after which the Governor would have his engineers and masons build it. Le Gentil could hardly believe this change in his fortunes: "with my soul content and satisfied I await with tranquillity until the approaching ecliptic conjunction of Venus with the sun comes to terminate my academic courses."

As a site for his observatory Le Gentil chose what he refers to as "the remains of a magnificent palace." It was partly in ruins, thanks to the British and their siege, but the base over a vault six to seven feet thick was still very solid. Here the observatory was built. True, it had one feature not very common to observatories: "the basement of my observatory served also, in the end, for over six weeks as a magazine for more than sixty thousand weight of powder. In spite of that, since [the Governor] had given me the liberty of inhabiting my observatory, this circumstance did not interrupt the course of my observations." Nor his serenity either, apparently.

Now, at his leisure, Le Gentil mounted and checked his instruments, and began the usual ancillary observations. Even the English over at Madras, as though to make amends for the misfortune they had earlier caused him, sent him an excellent achromatic telescope for his additional use.

What to do for the year he had still to wait? The industrious Le Gentil immediately threw himself into a detailed study of his surroundings, and in particular became absorbed in a study of Indian astronomy.

I amused myself also during my stay at Pondicherry in making some acquaintance with the astronomy, the religion, the habits, and the customs of the Indian Tamoults whom very improperly we call Malabars. What I had heard of their astronomy piqued my curiosity; but what finally sharpened it was the ease with which I saw calculated before me, by one of these Indians, an eclipse of the moon which I proposed to him, the first which occurred to me. This eclipse, with all the preliminary elements, took him only three quarters of an hour of work. I asked him to put me in a position to do likewise, and to give me every day an hour of his time. He consented to it; and when I asked him in how much time I could hope to be in a position to calculate an eclipse of the moon according to his method, he answered me, with an air rather indicative of conceit,

that with ability I would be able to do as much as he at the end of
six weeks.

This answer did not rebuff me, it only made me more curious.
I bound myself to take for about an hour every day my lesson in
Indian astronomy. Whether it was the fault of my master or whether
it was mine, or whether it was that of the interpreters (I changed
them three times), I needed more than a month of work at an
hour per day to be able to calculate an eclipse of the moon. . . .

Pursuits such as these (on which Le Gentil would one day write a
hefty volume) saw the intrepid astronomer through 1768 and into 1769.
Not only was he thoroughly enjoying this work, but, with the transit
steadily drawing nearer, he was pleased with the excellent weather he
was experiencing:

The nights at Pondicherry are of the greatest beauty in January and
February; you cannot have any idea of the beautiful sky which these
nights offer until you have seen them. . . . During the whole month
of May, until the third of June, the mornings were very beautiful. . . .

At Pondicherry the transit would take place in the early dawn hours of
June 4, and in the preceding days Le Gentil could hardly contain his
enthusiasm ("I was prepared. . . . I was awaiting the moment of the
observation with the greatest impatience.") The evening before was
wonderfully clear and calm and Le Gentil and the Governor—by now
evidently an astronomy buff—spent it in observing Jupiter's moons.
After nine years away from home, Le Gentil's great moment had finally
arrived; the anticipation was too much to allow of anything but a light
sleep:

Sunday the fourth, having awakened at two o'clock in the morning,
I heard the sand-bar moaning in the south-east; which made me
believe that the breeze was still from this direction. . . . I regarded
this as a good omen, because I knew that the wind from the south-
east is the broom of the coast and that it always brings serenity;
but curiosity having led me to get up a moment afterwards, I saw
with the greatest astonishment that the sky was covered everywhere, especially in the north and north-east, where it was bright-
ening; besides there was a profound calm. From that moment on I
felt doomed, I threw myself on my bed, without being able to close

my eyes. I no longer heard the bar in the south-east, but in the north-east; it was another very bad omen for me. Indeed, when I got up a second time I saw the same weather still, the north-east was even more overcast.

The dawn itself was heralded by a violent storm, quite uncharacteristic for the time of year:

> The sea was white with foam, and the air darkened by the eddies of sand and of dust which the force of the wind kept raising continually. This terrible squall lasted until about six o'clock. The wind died down, but the clouds remained. At three or four minutes before seven o'clock, almost the moment when Venus was to go off the sun, a light whiteness was seen in the sky which gave a suspicion of the position of the sun, nothing could be distinguished in the telescope. Little by little the winds passed . . . the clouds brightened, and the sun was seen quite brilliant; we did not cease to see it at all the rest of the day. . . .

Le Gentil's disappointment was so great that he could scarcely bring himself to comprehend what had happened:

> That is the fate which often awaits astronomers. I had gone more than ten thousand leagues [30,000 miles]; it seemed that I had crossed such a great expanse of seas, exiling myself from my native land, only to be the spectator of a fatal cloud which came to place itself before the sun at the precise moment of my observation, to carry off from me the fruits of my pains and of my fatigues. . . . I was unable to recover from my astonishment, I had difficulty in realizing that the transit of Venus was finally over. . . . At length I was more than two weeks in a singular dejection and almost did not have the courage to take up my pen to continue my journal; and several times it fell from my hands, when the moment came to report to France the fate of my operations. . . .

Some time later there came news that must have plunged him even deeper into gloom: at Manila the sky had been perfectly clear!

There was nothing to do but pack up and go home. But Le Gentil's travails were far from over. He had intended leaving Pondicherry soon after the transit, but now fell ill with a severe fever and dysentery that kept recurring, so that it was not until March of the following year that he embarked in some desperation for France. His convalescence was not improved by the voyage, and he had to break his journey at Isle de France once again in the hopes of regaining his health before continuing.

It was a slow process, not much helped by his witnessing the death of a compatriot astronomer by the name of Veron. They had had a short but pleasant acquaintance in Pondicherry, when Veron had called there, and subsequently Veron had been on a voyage to the South Seas with the great De Bougainville. Like Le Gentil in 1761 Veron had missed the transit by being at sea on the critical day. Here he was at Isle de France now, en route home.

> He arrived at Isle de France in extremity, from a fever which he had acquired by his great zeal to observe throughout the night on land when he was at the Moluccas; he died three or four days after disembarking from ship, July 1, 1770.

Le Gentil had some dealings with the island's commissioner over Veron's effects, and now, to Le Gentil's disbelief, the commissioner tried to induce him to undertake a voyage to the South Seas himself. Le Gentil could only speak of a disgust of travel beginning to lay hold of him; to get back to France was rapidly becoming an obsession.

His chance came in the form of one of the East India Company's ships which arrived at Isle de France a few weeks after Veron's death. Such was the languid way of things, however, that it was not until late November that the ship sailed, Le Gentil practically hopping with impatience at the unnecessary delays on the part of Officialdom. Surely he was at last rid of Isle de France for ever. But Le Gentil's luck was at least consistent, if nothing else:

> December 3rd [off the Isle de Bourbon] we were attacked by a hurricane which forced us to weigh anchor on the broadside and gain the open sea; it was then noon. Towards evening the try-sail was put out under the fore. During the night the violence of the wind and sea was so great that the helm of the rudder broke in its mortise; while the carpenters were busy repairing the helm, the tide-wave broke the bowsprit mast from its gammoning: this fall pulled down the main top-mast and the mast of the main top-gallant sail of the mizzen-top, which all came down in a single fall; our main-yard was badly damaged, and I regard it as a sort of a miracle that our main mast did not fall; for our main-shrouds had then more than six inches of slack; besides that, we were leaking in all

parts. We took six or seven days to get into shape to reach the Isle de France again; we arrived there January 1, 1771, to the great astonishment of all the colony, since the last thing they expected was to see us again.

Poor Le Gentil! He was beginning to wonder if he was not doomed to spend the rest of his life on this wretched island. And now, to add to his woes, the commissioner whose suggestion of going to the South Seas Le Gentil had so roundly scorned, began to make life even more difficult. Ships returning to France from the island were hardly plentiful, and yet whenever a French ship did arrive, the commissioner found some reason why Le Gentil could not be allowed to take passage on it.

More than three months dragged by before Le Gentil was rescued from his frustration by the arrival of a large Spanish warship bound for Cadiz. Influential friends put in a word with the captain, Don Joseph of Cordova, and Le Gentil was accepted as a passenger. He was almost overcome by the courtesy with which he was greeted, compared to what had been his more recent lot.

How impossible it is for me to find terms to depict the obliging air with which Don Joseph of Cordova received [me], and to describe the pleasure which it appeared to give him. . . . As if Don Joseph of Cordova wanted to command a great ship solely from the desire of seeing me more comfortable there. . . .

Naturally, with Le Gentil aboard, the ship ran into terrible weather while rounding the Cape of Good Hope (once, after all, known as the Cape of Storms), which left Le Gentil with his heart in his mouth lest he be transported yet again to Isle de France. "I told Don Joseph of Cordova of my worry about the bad weather we were having; he assured me that he would go back only as a last resort." But the Spaniards were superb seamen and brought their ship through with little difficulty, although Le Gentil was moved to say that "the sea was horrible, as I had never before seen it." Coming from Le Gentil, it must have been bad.

In the Atlantic they caught up with several of the French ships on which Le Gentil had been refused passage, and proceeded northwards as a group towards Europe. Le Gentil was invited to switch his passage to one of the French ships, an invitation he rather curtly declined in view of what had happened at Isle de France.

On June 24 they encountered a British ship, which caused a considerable stir of excitement. Back at Isle de France everyone had been agog
at what had seemed like another impending war between England on the
one hand and France and Spain on the other. And now, having been at
sea and out of touch for months, they were not at all sure whether this
might not be an enemy they had encountered. Although the British ship
took no evasive action, they stopped it and had its captain come aboard
to explain himself. Fearing that the English captain would try to trick
them into letting his vessel go unmolested by claiming there was no war,
Don Joseph started off by saying they had recently met with another
Spanish vessel that had told them of the war, and that the captain was
now their prisoner.

> The English captain appeared much surprised at the statement: he
> answered that he knew nothing of this news; that it was indeed true
> there had been a great deal of preparation in England; that they had
> armed; but that on his departure they were busy in disarming be-
> cause the differences which had arisen among the three powers . . .
> were settled. He was asked if he could give us some proof of what
> he said. He offered to show us the London *Gazette* for which he
> sent from his ship, and in which we saw the truth of what he had
> told us.

Don Joseph, as polite and cordial as ever, promptly produced wine and
biscuits and macaroons, and everyone drank to everyone else's bon voyage. The English captain, not to be outdone in the courtesies, immediately on returning to his ship sent over a large bag of potatoes and
"butter in proportion," a gift that Le Gentil rather sniffily summed up
in the phrase "At sea everything seems good. . . ."

The sea had one last fling at Le Gentil when contrary winds blew them
off course north of the Azores, but it cost them only a week, and then he
was in Cadiz. Like Pingré before him, he chose to make the rest of the
trip by land. On October 8, 1771, he too finally crossed the Pyrenees:
"At last I set foot on France at nine o'clock in the morning, after eleven
years, six months and thirteen days of absence." He didn't bother with
the hours and minutes.

But fate was not yet through with him. It had been so long since he
had last been heard from that he had meanwhile been presumed dead,
and he recorded his sardonic amusement that "people went to their

windows and doors when I passed through the streets, and I had many times the satisfaction of hearing people recognize me and attest loudly that I was really alive." This, however, had another side to it. His heirs and creditors were busy dividing up his estate, which had in any case been the victim of thievery, and the final blow came when he discovered that his place in the Academy had been awarded to someone else.

The last was rectified by the establishment of a special seat in the Academy, but it took a long court action to retrieve what was left of his estate, and Le Gentil had to pay the costs of this himself.

However, with that behind him, Le Gentil saw out the remaining twenty years of his life in relative serenity. Not too surprisingly, he gave up active astronomy and instead busied himself with writing his voluminous memoirs. He also married and became, it was said, a devoted husband and adoring father. In a way, perhaps, he managed to cheat fate in the end, for he died peacefully in October of 1792 (aged 67), just months before the worst horrors of the French Revolution descended upon Paris. Still, Cassini in writing that Eulogy said of him: "His face did not prejudice one in his favour . . . [and] in his sea voyages he had contracted a little unsociability and brusqueness. . . ." As well he might.

What of the results of all these enormous efforts? Did Mason, Wales, Chappe, Le Gentil, and all the many others achieve what they set out to do, which was to determine the distance of the sun from the earth with the greatest accuracy? The answer is both yes and no.

Initial calculations based on their observations gave a distance to the sun of about 95 million miles, although, curiously, no single figure representing the "best" answer would emerge for many, many decades. This was because observations from any two sites could be used to derive an answer, and thus there were as many answers as there were pairs of sites. All told, the observers had in a way succeeded too well: they had provided more information than was actually needed. So it happened that everyone had his own favourite pair of observations, or felt that certain pairs should be ignored for one reason or another, while others should be given preference, and the result was that many different answers for the sun's distance emerged. It would not be until the nineteenth century that the mathematician Karl Friedrich Gauss would find a method of combining an excess of data to yield a single "best" answer.

But that difficulty apart, it would still be many years before the observations could be made to yield their full potential. This happened because a major uncertainty in the early calculations was the precise latitude and longitude of each observer. The latter had done their best to determine these, but the methods (particularly for longitude) of the eighteenth century were still crude. Later, in the nineteenth century, when the locations of the observers could be determined more accurately, their actual transit observations could be re-used to obtain a more accurate result of close to 93 million miles. As such, then, the eighteenth-century efforts must be counted not merely as worthwhile, but as a downright success.

On the other hand, the method never quite lived up to the promise that Halley and Delisle expected of it. And the reason for this was that curious phenomenon remarked on by James Cook when he described his observation of the 1769 transit: "We very distinctly saw an Atmosphere or dusky shade round the body of the Planet which very much disturbed the times of the Contacts. . . ." This effect was later to be called "the black-drop effect"; it is caused, just as Cook said, by Venus having a particularly dense atmosphere which refracts the sunlight coming from behind and creates an optical illusion. By contact the astronomers meant the exact instant when the edge of the planet touched the edge of the sun. What they saw, however, was that as the planet began to encroach on the disk of the sun, the edge of the planet seemed to be drawn out into a black drop attached to the edge of the sun, and it was not until the planet was obviously well into the sun's disk that the two edges separated again. Likewise, at the end of the transit, when Venus was approaching the opposite edge of the sun, the two edges would appear to meet even while the planet clearly had some distance to go. Thus it was impossible to determine the times of contact with the precision that Halley had thought possible, observers at the same site often differing by many seconds in their estimates.

Nevertheless, the transits of Venus remained the best available means for determining the sun's distance for more than another century. When the transits of 1874 and 1882 came around, there were again many expeditions sent out (although with considerably fewer adventures!) in the hopes of obtaining a definitive answer. By then astronomers had photography as a tool, and they hoped that by photographing the planet at every step of its way across the sun it would be possible to avoid the worst of the black-drop effect. Once again the results did not quite live up to expectations, and although an improved answer for the distance of the sun was obtained, it was clear that the usefulness of the transits of Venus had reached a limit. The method rapidly gave way to other techniques developed around the turn of the twentieth century, and they in turn have now given way to the methods of radar astronomy. Today we know the distance of the sun to a precision of a few miles, the equivalent of measuring six miles with an uncertainty of less than a single human-hairbreadth!

But at the very least, when 2004 again brings the quaint novelty of another transit of Venus, we may raise a silent toast to the transit observers of the eighteenth century.

# "...IN THE COUNTRY OF THE HOTTENTOTS"

*Maclear and Herschel in South Africa*

The mountain dominates all. Almost four thousand feet high, its start-lingly flat top sweeps east and west for miles, and the grey granitic slabs of its face hang precipitously over the city below. To the east, and slightly in front, is the skewed hulk of Devil's Peak, while to the west and also in front, the fingered peak of Lion's Head sweeps northward down to Lion's Rump. Between them they cradle the heart of Cape Town.

Table Bay, tavern of the seas to European ships for four hundred years, laps against the city at the mountain's foot, and its wide white beaches run on virtually unbroken for hundreds of miles northwards up the coast.

A road winds out of the city around the Rump and down the west coast of the Peninsula, passing under the western buttress of the moun-tain known as the Twelve Apostles, through the sleepy little fishing village of Hout Bay, and then, as one of the most spectacular of roads anywhere, it is carved across the thousand-foot face of Chapman's Peak where the latter hangs vertically above the sea. From there it runs on southwards through the more desolate regions of the Peninsula, where the drowned kelp-beds wave ceaselessly in the cold waters of the Atlantic.

But if one travels the Peninsula along the eastern route, a different panorama unfolds. From the lower slopes of Devil's Peak the view through the tall pines is across the wide expanse of the Cape Flats to-wards the wine-growing district of Paarl, nestled at the foot of the Hottentots-Holland mountains, whose dragon-like spine dominates the eastern skyline fifty miles away. The road winds on through the rain-forest of Kirstenbosch alongside the eastern escarpment of Table Moun-tain, and down through the beautiful Constantia valley to where the warm Indian Ocean washes the eastern coast of the Peninsula. Continu-ing on past the naval base of Simonstown, the east coast road finally merges with its western counterpart to bring the traveller to Cape Point, the southernmost tip of the Cape Peninsula. A knife-like wall of rock four hundred feet tall, it runs out to divide the two oceans. Here, legend has it, the Flying Dutchman ceaselessly patrols the seas, doomed for his blasphemous curse that he would round the Cape if it took till eternity.

Standing on Cape Point, however, one is more likely to see a rocky promontory that juts out into the Atlantic a mile or two away. It is slightly bifurcated. The eastern segment is the actual geographic Cape of Good Hope. The western segment is Cape Maclear.

✦✦✦

*Aerial view of the Cape Peninsula , South Africa. (This appeared in the Sep-
tember 1975 issue of South African* Scope *and is reproduced by courtesy of
the South African Embassy in Ottawa.)*

The sunny summer's morning of January 5, 1834. Thomas Maclear, a short, dapper man of thirty-nine, his curly hair ruffled by the south-easterly breeze, is standing on the deck of the *Tam O'Shanter* as it bobs gently in Table Bay. He is the new director of the Royal Cape Observatory, and as he looks across to the great mountain with the little town below, he has very mixed feelings.

On the one hand is the immense relief of having arrived at his destination after a hideous sea-voyage of several months. But on the other there is his new job. For one thing it bodes ill that he is already the third director in hardly more than a dozen years. For another there are the instructions as to his duties: he is to compile a catalogue of the southern stars, giving special attention to a selected list of standards; he is to observe the fundamental stars of the European observatories in order to carry out special refraction studies; he is to make an uninterrupted series of observations of the sun, moon, and planets; he is to observe lunar transits and moon-culminating stars; he is to refine the longitude of the Cape; he is to make regular meteorological observations; he is to observe the tides and establish a standard reference level; he is to lay plans for and begin work on the measurement of a geodetic arc of meridian; he is to comply with requests from His Majesty's ships for the regulation and rating of their chronometers; and, in case there is anything not covered by the foregoing, he is "to advance astronomy, improve navigation, and promote the credit of the observatory." This might well be a task for a small army of workers, but Maclear is to carry it all out single-handed with the aid of only one ill-trained assistant. Furthermore, he is in a primitive and in many ways savage land, so when, inevitably, things go wrong, self-reliance will be the order of the day. Experience? Maclear has none. Until this appointment he was not even a professional astronomer; he was a medical doctor in the quiet countryside of England.

Maclear's instructions reflect the early purposes of the Cape Observatory. Britain had acquired the Cape less than twenty years earlier as a consequence of the Napoleonic Wars. Not that Britain viewed her new colony with much enthusiasm, and the existing Dutch colonists there certainly viewed the take-over with no enthusiasm at all; it was merely that the Cape, commanding the great sea route to the East, was just too strategically valuable to be left to the mercy of any possible French invasion. So the British got in first.

Very soon after that the British Admiralty began to concern itself increasingly with the problems of navigation in the southern oceans, now that its fleets would show a more or less permanent presence there. Marine navigation had been enormously advanced with the invention of the chronometer half a century before, but it was still necessary to have accurate star catalogues, and the stars of the far southern skies are inaccessible to northern observatories. The only way to obtain accurate star positions for navigation by southern stars would be to have a southern observatory.

So it was that within six years of the British taking over the Cape once and for all (it had been an off-again on-again affair for twenty years before that), there was talk of establishing an observatory there that would be the southern counterpart of the Royal Observatory at Greenwich. In the end it would survive there under the aegis of the Admiralty for almost 150 years.

Everyone agreed it was a great idea, and the Order-in-Council establishing the observatory was signed on October 20, 1820. That was easy. Finding an astronomer foolhardy enough actually to go and make the place work was something else. But such a man was forthcoming: the Reverend Fearon Fallows, a Fellow of St John's College, Cambridge, of whom a contemporary once remarked "It is difficult to conceive of a man of such simplicity of character and such absence of knowledge of the world in the nineteenth century." He would go.

Fallows and his wife arrived in Cape Town in 1821 and began work on the new observatory. The first task was to select a site, the criteria for which were that it had to be government-owned land, had to be far enough away from the mountain to have unobstructed skies, yet close enough to Table Bay for visual time-signals to be sent to ships anchored there. Only one spot seemed to meet all these requirements, a low hill a few miles out on the Flats from Devil's Peak which gloried in the name of Slangkop, meaning "snakehill." The name was accurate, as several astronomers would later testify in unequivocal terms. Additionally it was almost devoid of soil, while being surrounded by extensive marshes, down to which a variety of wild animals would occasionally make their way. For years, in fact, there existed a body of folk-lore on the conduct of astronomy in the presence of various unsavoury beasts. There was the hippopotamus that wandered down and became stuck fast in the mud, its irate bellows causing the tiny staff of the observatory to wonder how to free a couple of tons of exceedingly angry and dangerous hippo. Finding the engineering requirements beyond them, they decided to dispatch the wretched animal instead, but now discovered that the hippo's hide was impervious to their available guns. It took a neighbouring burgher to show them that one must first hack a hole in the hide and then fire through that to kill the animal. Not a task one really expects when deciding to become an astronomer!

There was also the case of the astronomer who went to open the roof above his telescope preparatory to observing early one evening, and found it apparently stuck. Climbing up the outside of the building to free the mechanism, he found his surmise mistaken: it was a leopard sleeping across the metal, still warm from the late-afternoon sun.

But in 1821 the Rev. Fallows' concerns were mainly with constructing a building that would serve both as observatory and home. It was a slow and tedious task, beset with the difficulties of a neophyte in a strange and primitive land trying to find contractors to do the work properly. In fact it was not fully completed until 1827, but, in mute testimony to Fallows' perseverence, it stands to this day as the observatory's main building.

Meanwhile Fallows began observing with a portable telescope, while, he noted none too happily, the jackals howled dismally in the darkness around him. His days, when not spent on astronomical work or keeping after the builders, were devoted to improving the site. To this end he opened a school for the children of neighbouring farmers, the cost of education being one wagon-load of soil per lesson. With soil now available, he began to think of planting out the grounds. In particular he found the strong south-easterly trade winds (known locally as the Cape Doctor for their scouring of refuse from the streets) bothersome, and applied to the Admiralty for money to plant trees as a windbreak. He was stiffly refused, with a rebuke that if he wished to beautify the observatory it would have to be at his own expense.

Fallows had to do almost everything himself, from gardening to astronomical observing. He found it impossible to keep an assistant, and often relied on his wife to stay up through the night to help with the recording of his observations. Slowly the work and worry took their toll, and when he contracted scarlet fever in 1831 he failed to recover, dying at the age of 43. His grave is on the sward before the observatory's main building, shielded by the trees the Admiralty had once refused him.

Fallows' successor as director of the Cape Observatory was a remarkable man by the name of Thomas Henderson. Born in Dundee, Scotland, in 1798, Henderson was a fully trained lawyer, being at one stage advocate clerk to a judge of the Supreme Court of Scotland. Throughout his twenties, however, he spent all his spare time studying astronomy, attended scientific meetings whenever he could, and with the publication of one or two outstanding papers, soon garnered a reputation as an excellent astronomer. By the time he was thirty he was being highly touted for the Chair of Practical Astronomy at Edinburgh University, a position eventually lost to him through the procrastination of the authorities for several years. For a while it looked as though he would take over the running of the Nautical Almanac Office, but when the Admiralty were suddenly faced with the need to replace Fallows, they instead offered that job to Henderson. He was not at all sure he wanted it, thinking Africa would be a far cry from his beloved Scotland, and only some very persuasive talking on the part of his friends overcame his reluctance.

He arrived at the Cape in April of 1832, and triumphantly announced that his worst fears were fully confirmed. He took an instant loathing to the observatory, and for the rest of his life always referred to it as "Dismal Swamp." Later, safe in the security of Scotland again, he wrote to Maclear: "I will tell you about my residence in Dismal Swamp among slaves and savages—plenty of insidious venomous snakes. What would you think if, on putting out your candle to step into bed you were to find one lurking beside the bed?" Maclear was soon to find out what he would think.

Apart from these hazards, Henderson found that his greatest problems lay with his superiors in London. Having dispatched their man to this outpost of Empire, they were little interested in his activities, and expected him to get on with the job as best he could. Henderson kept up a running battle with them, trying to obtain financial support for improvements to the observatory, but, like Fallows before him, met with little in the way of positive response. One of his much later successors, Sir David Gill, wrote of him: "Henderson was rather the refined observer than the pioneer. . . . He was not the man to fight an uphill battle with neglect at home, and to compel Fate, in the shape of official indifference or incapacity, to do his bidding. . . . It was impossible for the strongest to adequately fulfil the duties of his office without more assistance; he saw that the situation, as it stood, was an impossible one, and he was too honourable a man to accept the emoluments of an office without the most punctilious discharge of his duties. Accordingly, in May 1833, he resigned his post. . . ." To read Gill you'd think Henderson was sorry to go. Neither does Gill mention the final straw that precipitated the resignation: a fight with the bureaucrats over the architecture of the observatory's sanitary arrangements. Surely one of the less common reasons that have caused directors to resign their posts.

But Gill was quite right about the punctilious discharge of duties. Henderson worked like a dog. During his one brief year at the Cape he accumulated vast piles of observational data, and when he happily went back to Scotland (to become Astronomer Royal for Scotland), he had years of hard work still ahead of him in analysing his results. Here was his misfortune.

Ever since the beginnings of modern astronomy four hundred years earlier, astronomers had made the most strenuous efforts to measure the distance of at least one star. To no avail. By the nineteenth century it had become a bitter irony that the most fundamental thing—the distance —about astronomy's most fundamental objects—stars—was as yet completely unknown. Here, for example, is what the President of the Royal Astronomical Society in England had to say about it:

> To accomplish this has been the object of every astronomer's highest aspirations ever since sidereal astronomy acquired any degree of precision. But it has been an object which, like the fleeting fires that dazzle and mislead the benighted wanderer, has seemed to suffer the semblance of an approach only to elude his seizure when apparently just within his grasp, continually hovering just beyond the limits of his distinct apprehension, and so leading him on in hopeless, endless, and exhausting pursuit.

In short, finding the distance of a star was astronomy's greatest problem.

What Henderson didn't know was that buried among his unanalysed data was the inadvertent determination of a star's distance. He had been back in the Highland mists a full five years before he realized what he possessed, but by then it was too late. Only months before, two of Europe's most famous astronomers, Wilhelm Bessel and Wilhelm Struve, had achieved the long-sought goal. Henderson had to settle for third place. Whether it was discouragement over this, or whether the snakes and savages had taken a heavier toll than believed, Henderson did little else with his life, and died only a few years later at the age of 46. Almost nothing is known about him as a person. Although he is mentioned in all the textbooks as the third of three discoverers of stellar distance, we lack even a portrait of him. He is one of astronomy's least-known famous men.

And now here was the new director, Thomas Maclear. He had been born in Ireland in 1794, and, like many an astronomer before him, had seemed destined by the fond wishes of his parents for a respectable career in the Church. But he rebelled against this, and after the age of fifteen was steered into a medical career by two eminent uncles, themselves doctors. His first appointment was as House Surgeon at the Infirmary in Bedford, England, and it was while there that he first became friendly with a remarkable family of Smyths.

Admiral Smyth had enjoyed considerable attention while commanding a British fleet in the Mediterranean during the Napoleonic Wars, but, a true eccentric at heart, his real passions lay in archaeology and astronomy. He published several very odd books on these subjects, although at least one of them, *A Cycle of Celestial Objects,* achieved considerable popularity. At home in Bedford he maintained a formidable arsenal of astronomical instruments, and with these the entire family seem to have indulged themselves to an extreme. So enthused over astronomy was Admiral Smyth that when he encountered the Italian astronomer Giuseppe Piazzi during a stay in Sicily, he named one of his sons Charles Piazzi Smyth. (This son would, as a teenager, become an assistant to Maclear in South Africa, and later in the century rose to the heights of Astronomer Royal for Scotland. Eventually his eccentricity exceeded even that of his father, when he became convinced that the destiny of the human race was bound up in the dimensions of the Great Pyramid. Indeed, all the Smyth children were talented, if not eccentric. One son became a professor, another an army general, while one of the daughters became the mother of Lord Baden-Powell of Boy Scout fame.)

To befriend a Smyth was to befriend astronomy, and soon, like the young lawyer Henderson elsewhere, Maclear was spending all his spare time studying the subject. He took up a private practice in the town of Biggleswade some miles away, which was both lucrative and demanding, but his obsessional interest in astronomy, flames fanned by the Smyths, continued. He began attending meetings of astronomers, and within a few years had come to be accepted as a fully-fledged member of the profession. So it was that in 1833 the Admiralty, undergoing its customary difficulty in finding an experienced astronomer to take the job, offered the directorship of the Cape Observatory to Maclear.

The avuncular eye viewed this development with no small alarm. Astronomy might be suitable as a hobby for the mildly eccentric, but it was hardly a profession fit for a gentleman. And as though that weren't bad enough, Maclear was proposing to take his wife and young family off to what the Encyclopaedia Brittanica referred to as "a promontory of Africa, where the Dutch have built a good town and fort. It is situated in the country of the Hottentots." That authoritative work further warned of the nature of the inhabitants: "The ladies are lively, good natured and familiar: and from a peculiarly gay turn, they admit of liberties that would be thought reprehensible in England. . . ." Certainly not the place for a respectable young family man.

But Maclear was not to be dissuaded by his uncles. His major problem was finding the money to transport his family the six thousand miles to Cape Town. One might have thought that the Admiralty, of all bodies, could have easily enough made arrangements for getting their astronomer to his post; but no, Maclear would have to pay his own way.

A tentative approach to his uncles soon confirmed that not a farthing would be forthcoming from that source. Yet for the family to sail in the relative comfort of an East India Company ship would cost no less than £350—the equivalent of many thousands of dollars in terms of today's purchasing power.

They would have to go by a Government ship after all. Francis Beaufort, the Navy Hydrographer and an acquaintance of Maclear, finally managed to arrange passage for them on the *Tam O'Shanter* at what appeared to be the remarkably attractive price of only £36. Prudently, they had a friend observe the *Tam O'Shanter*'s captain as he sat drinking in a tavern, and although he passed this test, it proved in the end to have been insufficiently stringent.

Meanwhile Beaufort wrote to invite Maclear "to come up on Monday to see your cabins and order your pigs and chickens and potted meats." It would also be necessary for Maclear to lay in a good store of well dried potatoes and apples as antiscorbutics. In fact a good deal of attention had to be paid to their shipboard supplies, because for £36 the Government ship would supply no more than the most basic of rations, and as Maclear wrote to his wife, "At all pecuniary risk I am determined to save you from salt beef and sour joints." (The correspondence between Maclear and his wife Mary has some novel twists. She, in the manner of early nineteenth-century wives, always addressed him as "My Dear Maclear," while he on occasion would unbend to address her as "Coddles.")

Finally, the two Maclears, their five children, two servants, and a governess, sailed aboard the *Tam O'Shanter* at eight o'clock on the morning of October 10, 1833. For the two principals, at least, it was farewell to Britain as home for ever.

It was a ghastly voyage. They were all horribly seasick most of the time, and despite all the careful preparations, one of the children died. This later brought a lugubrious letter of consolation from Mrs Smyth, saying she had never considered the child to be strong anyway.

And just to add to Maclear's woes, the captain announced that according to his reckoning Maclear owed him a further £220 for the passage, a claim whose outcome is unfortunately unknown.

Undoubtedly, then, the family must have been eager to be ashore on that January morning of 1834, even if it only meant heading for—as Gill puts it—"the still desolate looking observatory".

Waiting on the dock for them was Maclear's new assistant, Meadows by name, who had been looking after things since Henderson's departure, and who now came forth with one of the most remarkable salutations on record: "So, Sir, you have determined to accept this wretched appointment!" From this beginning Maclear and Meadows never looked back. Maclear's opinion of his assistant can be gleaned from a letter Maclear wrote to the former director: "Henderson, I know you well. . . . Those into whose hands you as a single man fell were without exception the most melancholic croaking helpless couple I ever met with. . . . [Meadows shows] a want of enthusiasm in a science the outlines of which he seems but slightly acquainted with."

In fact this reference to a "couple" had soon to be revised, for the Maclears rapidly became aware that Meadows was maintaining a *ménage à trois,* a certain Mrs Lee being a remarkably intimate member of the Meadows household. When Maclear protested strongly against this scandal, demanding that Mrs Lee be sent packing forthwith, he was astonished to find Mrs Meadows reduced to tears, pleading with him that the other woman be allowed to remain at least until a passage to Van Dieman's Land could be arranged for her.

Meadows was a good judge of character, it seems. He managed to push Maclear to just the point where Maclear had made up his mind to fire him, whereupon he forestalled Maclear by blandly resigning on the grounds of ill-health, and went back to England to become Secretary to a Gas Company.

Assistants, as Fallows could have told Maclear, were hard to come by. Eventually Maclear decided to appoint his own manservant, Thomas Bowler, as such. Bowler was "sober and honest and an excellent penman," and would henceforth rejoice in the title of "Labourer," a post that covered more or less everything from gardening to running errands to carrying out astronomical work. There would be a pay increase to £70 a year, and a room in the assistant's wing of the observatory, where, Maclear noted, Bowler might "boil his own kettle morning and evening and cook his own steak. He can be as comfortable at the Observatory as anywhere. The butcher and baker come every second day." This was luck, for prices in Cape Town itself were high: "A 10 lb fish is 2d or 3d, butcher's meat 1½d to 2d per lb. Bread, vegetables and butter as dear or dearer as in England. Bowler is steady and if he can acquire an obliging disposition to his employer he will do well."

Alas, Bowler proved insensible to these advantages. For one thing it turned out, thanks to Fallows' architecture, that Bowler's room was traversed each evening by the astronomer on his way to and from the telescope, which, as Bowler plaintively noted, left him no privacy should he "chuse to marry." Also, the boiling of the kettle and cooking of the steak were not much to his liking, and before long he preferred to take his meals in the village of Rondebosch a few miles away. This meant that he would be absent from the observatory for something like a couple of hours each mealtime, which very soon brought him into conflict with Maclear.

Their relationship rapidly deteriorated and came to a sharp end when Bowler insulted Mrs Maclear in some (unrecorded) way. Maclear, hardly able to "command his feelings," immediately sent Bowler packing. Subsequently, Maclear was again hardly able to command his feelings when he heard that Bowler was contemplating taking legal action against him, but Bowler apparently thought better of it and nothing came of it.

There is, however, a neatly ironic twist to Bowler's story. When Maclear said of him that he was an excellent penman he had not been exaggerating. Although no genius, Bowler did have quite remarkable talents at sketching and drawing, and after leaving the observatory he later became an art teacher at the Diocesan College in Cape Town. Here he

*Portrait of Thomas Maclear*
*(by courtesy of* **H.M.** *Astronomer at the Cape)*

remained for over thirty years, during which time he produced an abundance of sketches of the town and its environs. As it happened, he was almost alone in this task, so that much of modern knowledge of what Cape Town looked like in the mid-nineteenth century is due to Bowler's work. Today a Bowler sketch carries a considerable price, and the name is well-known in South African art circles. Few have ever heard of Maclear. Yet when Bowler finally left the Cape in 1866 it was to die penniless three years later.

Early 1835 brought Maclear one of his most unpleasant escapades: the affair of the duel. It arose in part through his own autocratic nature, for he was a peppery little man, and not for nothing known to his underlings as "The Emperor."

He had been up observing all of one February night, and so next morning he naturally slept very late. He was, in fact, awakened by a servant with the news that a group of French naval officers had come to visit the observatory and were awaiting him below. The demands of nineteenth-century courtesy being what they were, Maclear scrambled out of bed and asked the servant to show the visitors into the library while he dressed. Living quarters and working quarters at the observatory were very much juxtaposed, as Bowler had noted, and Maclear had fears that his visitors would wander into the so-called transit room where some of his precious and very delicate instruments might become objects of curiosity to their uneducated hands. The servant therefore had strict instructions that the visitors were to remain in the library until Maclear arrived.

On coming down, however, Maclear found the library empty and, sure enough, to his extreme anger the party had adjourned to the transit room. He gives the Englishman's understated account of proceedings: "I enquired if my servant had not delivered a message to which they gave no direct answer, however, and went off." Behind that one can imagine a good deal of expostulation on the part of the fiery Director.

Within the hour two of the group were back, to hand to the astounded Maclear the following letter:

                                                                23rd Feb 1835
      Sir!
            As I have been at the Royal Observatory being a lover of the science & being insulted by your appearance    If it is your intention by so doing and as you are Officer of the navy and I Commander of

the French man of war the Madagascar at present in Table Bay    I
wish to have satisfaction for your improper Conduct, & request for
an answer
>I am
>Bosse
>Commander of the French Ship Madagascar

This "bombastic note," Maclear observed, was "written on a slip of dirty
paper and sealed with a bit of wax with the impression of the thumb."
Having read it, "I asked the gentlemen who conveyed it for their cards—
they had none—I asked them to furnish me with their names. After some
hesitation one gave his name, Truter of Cape Town, the other J. Bestandig. I enquired if Mr Truter was the son of the neighbouring Baker, 'No,'
said he, 'I am a cousin of his.' (On enquiry I found he told a lie. He is the
son of the Baker Truter and follows the same Trade himself.)"

What seems to have incensed the Emperor as much as anything was the
fact that Bosse had chosen mere tradespeople as his seconds, and that
they had come without even a gentleman's accoutrement of visiting cards.

However, a certain amount of conciliation seemed in order, so Maclear
sat down and penned the following reply:

Royal Observatory, Feb 23, 1835

Sir,
>I have the honour to inform you that I am extremely surprised
at the nature and tone of a note received from you at this moment. I
disclaim all intention of insulting you or any man. I beg that you will
distinctly understand that the Royal Observatory is purely a scientific establishment—the Instruments are of a most delicate construction and in order to avoid any derangement of them I follow the Rule
of other and similar scientific Institutions, viz. that no visitor
shall be allowed to inspect them unless he is accompanied by myself or some one deputed by me for the purpose. I am responsible
to the Crown for their safe custody, while my character is involved
with the observations that are made with them.
>When you honoured the Observatory with a visit this morning
I desired my servant to show you and those who accompanied you
into the Library until I could come downstairs, having been up all
night in the exercise of my profession and I was much annoyed to
find you and those with you in the Transit Room examining the Instruments. I disclaim having employed improper language on the
occasion or anything insulting.
>I believe it is well known that I never hesitate to show the Ob-

servatory to any proper person who makes application for the purpose and particularly foreigners and I can only say I shall have
pleasure in doing so to you,
>    I have the honour to be Sir,
>    Your obedient servant
>    Thos. Maclear

To M. Bosse,
Commander of the French ship Madagascar.

But he was by no means sure that this would close the matter. He sent
copies of the two letters to Captain Bance, a senior officer over at the
naval base in Simonstown, "stating the circumstances and requesting his
opinion on the following points—If he considered my honour in any way
involved? If so I would go out with this Bosse, provided he was a com-
missioned officer and his friend neither a Baker nor a common lodging
house keeper (which I understand from my servant this Bestandig is)
but my family and my insurance were barriers to fighting unless my
honour was concerned, in that case nothing should prevent me."

There is a certain practicality to this. Somehow, from the viewpoint
of the present one tends to look back to the days of duelling with a romantic eye that doesn't include considerations of the combatants' insurance
policies.

Happily, Commander Bosse appears to have let the whole matter drop
at this point; no more was heard from him. The naval authorities at
Simonstown, however, didn't let Maclear off nearly so easily. Captain
Bance came over to tell him that his conduct had been highly improper,
that "the note was more the production of a prize fighter than a gentleman
and that I ought to have sent it to the French Minister of Marine and
have not replied to him."

Through all these trials and tribulations Maclear had available to him at
least one firm friend and advisor, a man only a couple of years older than
Maclear, and living only a few miles away, yet already one of the world's
most talented and famous astronomers: Sir John Frederick William
Herschel.

John Herschel was the only child of an even more famous astronomer,
Sir William Herschel. The latter had emigrated to England from Germany
in the mid-eighteenth century, being at that time a professional musician.

However, he devoted much of his spare time to astronomy, and in 1781 he had the superb good fortune to discover the planet Uranus. This event completely revolutionized his life; it brought him, of course, great fame, as well as an appointment as personal astronomer to the King, which enabled him to give up music as a profession and instead turn all his energies to astronomy. It had been mostly luck that had brought him to prominence, but his great natural abilities soon became evident. He became a leading builder of telescopes, a sideline he seems to have developed into an extremely lucrative business. But he showed himself very capable of using telescopes as well, and it is probably no exaggeration to say that he became the greatest observational astronomer of his day. Certainly, by the time of his death in 1822, his many awards and medals showed the enormous esteem with which his contemporaries regarded him.

William was already in his fifties when he married, so there was a larger than usual generation gap between John and his father. The boy had a lonely upbringing, with little apparent warmth between him and his parents. He seems to have had a much happier relationship with his Aunt Caroline, William's devoted sister. Caroline was a remarkable person who gave up a singing career to become William's housekeeper and indefatigable assistant during his bachelor days. So assiduous were her efforts at the telescope that she eventually became a first-rate astronomer herself, and in the days before the Royal Astronomical Society admitted women, she was one of only two or three women who received special recognition by that society. She outlived both William and his wife, returning eventually to live in her native Germany, and always remained the special confidante of John and, later, John's wife, Margaret. As the years went by and Caroline advanced in age, the Herschels seem to have lived in lugubrious and almost daily expectation of her demise. They should have known better. John visited her in Hanover when she was 83, and reported "She runs about the town with me, and skips up her two flights of stairs. In the morning until eleven or twelve she is dull and weary, but as the day advances she gains life, and is quite 'fresh and funny' at ten o'clock P.M., and sings old rhymes, nay even dances! to the great delight of all who see her." The fine old lady continued undiminished until she died at 98 in 1848.

John's early career stands in sharp contrast to his father's. Where William had been poorly educated and almost literally penniless when he arrived in England to make his own way in a strange country, John had

*Portrait of John Herschel by Henry William Pickersgill. Margaret Herschel thought this portrait an "admirable likeness." (By permission of the Master and Fellows of St. John's College, Cambridge)*

all the advantages available to the scion of a wealthy and famous family. His own abilities in science were evident as soon as he reached Cambridge, where he in due course stood first in the extremely difficult Mathematical Tripos. But his talents ranged far beyond this. Sophisticated, good-looking, commanding a witty turn of phrase in speech and writing, fluent in several languages, widely educated in the classics as well as in science, he was a young lion in Society.

He did not want to be an astronomer. Neither did his father want him to be one. William wanted him to enter the Church, but John declined (there seems almost an inevitability to this step among astronomers of that day), and went off to study law instead. But he gave that up within two years, and in 1815, still only twenty-three years old, applied for the Chair of Chemistry at Cambridge, a position he lost by only one vote. Indeed, if there was one single subject that John Herschel enjoyed throughout his life, it was chemistry. He made considerable contributions to the chemistry of photography, discovering the action of "hypo" in "fixing" photographs, inventing the terms "positive" and "negative," taking the first pictures on glass plates (of his father's biggest telescope), as well as making other discoveries.

That John eventually became an astronomer, it is clear from his writings, was purely out of a sense of filial duty. His ageing father had left incomplete a number of vast observational programs, and these John decided he must finish. He worked hard and industriously at it, and his own brilliance at astronomy soon brought him medals and prizes from such institutions as the Royal Society and the French Academy of Sciences; he would soon gain recognition as one of the world's leading astronomers, become President of the Royal Astronomical Society, etc, etc; yet his heart was never really in it. He would always lack that singular total commitment to the subject that was the hallmark of his father, and his diaries contain such entries as "Sick of star-gazing—mean to break the telescope and melt the mirrors."

In fact, John Herschel was never a professional astronomer in the fullest sense of the word. True, he had had a very professional training, and his research work was outstanding and of fundamental importance, yet he would never hold any position in a university or scientific establishment, would never earn his living from astronomy, would always be the wealthy, private individual.

By the early 1830s he was already famous, already knighted, and happily married to Margaret Brodie Stewart. He and "dear Maggie" would always enjoy as warm a relationship with each other as with the twelve children they eventually produced, a family life in sharp contrast to that of Herschel's own lonely childhood.

One of William Herschel's great projects had been to survey the skies for objects such as double stars, star clusters, and nebulae. But in England he had been restricted to the northern skies. John now decided to extend this work into the south, and to do so he took his family for a spell of several years to the new colony of the Cape in South Africa.

It was pure coincidence that his arrival there fell within a few days of that of Maclear. The two were already quite friendly from several visits by Maclear to the Herschels' family home in Slough, and when the Maclears had initially hoped to travel out by East Indiaman it was so that they might go on the same ship as the Herschels. The latter, of course, had none of the financial worries that beset Maclear.

The Herschel party embarked on the 611-ton *Mountstuart Elphinstone* on November 13, 1833. There was John and Margaret, the three children they then had, Carry, Bella, and William by name—William being less than a year old—a nursemaid, Mrs Nanson, and two men servants, one of whom (Stone) was Herschel's assistant and mechanic. The family itself occupied three cabins, one of them being no more than 6' by 7' in size, although Maggie happily noted "we have the best cabins on board, & in this snug little deck cabin of Herschel, I spend the whole day by his gracious permission. . . ." It would be a two-month voyage.

Herschel was an indefatigable recorder of events, and a never-ending stream of letters, memoranda, observations, notes, and diaries issued from his pen. Diary-writing in particular seems to have been a serious matter, witness the stern injunction on the flyleaf of one volume:

> This book belongs to J. F. W. Herschel
> Slough Bucks
> If lost, whoever may find it will receive five shillings reward on re-
> turning it to the above address. . . . No more reward will be offered
> nor will the book be advertised for, not to encourage dishonest find-
> ers The Memoranda are purposely made unintelligible and value-
> less to anybody but the owner.

The diary gives us a blow-by-blow account of their time at sea. Almost immediately they found the crew not to their liking, and Maggie proved to be a very poor sailor.

Saturday, November 16, 1833
The wind Rising and a considerable ripple coming on the water, both increased till at length it blew hard & the Capn was obliged to take in Top Gallants and put 2 or 3 reefs in main topsails an operation accomplished slowly & with difficulty, by reason of the extreme *inefficiency* or *insufficiency* or both of the crew, who are declared as bad a set of lubbers as ever worked a ship.—Sea grew at last very heavy & we passed a dreadful night. M[aggie] very ill indeed.

Monday, November 18, 1833
After a comfortless night filled with strange, connected, tragical dreams in my own little awning cabin found M. still extremely ill with sickness & headache but able to eat some meat.

Tuesday, November 19, 1833
5 of the sheep died since yesterday—probably from over-drinking as they have been allowed 2 quarts of water a day each (Sheep drink little & seldom) On the other hand the ducks of which there is a great collection on board get little or no water!
M. rose & . . . seems to have got over the worst of the sickness.

Herschel himself was made of sterner stuff, and his ever-inquisitive mind roamed restlessly for new things to do. The diary is filled with endless observations: continual readings of air and sea temperatures, wind conditions, descriptions of bird life, Herschel's latitudes and longitudes compared to those of the ship's officers, sky conditions at night, the excitement of first seeing the Magellanic Clouds (galaxies) in the southern skies, and, when all else failed, the maximum and minimum readings of his own pulse (53 and 47 beats to the minute). He measured the heat of the sun on deck, dropped white-coloured weights overboard to study the clarity of the water as they slowly sank, observed the flow of water around objects towed aft.

Herschel never felt restricted to the physical sciences:

Saturday, December 7, 1833
—A Dolphin caught. His back fin in dying grew a very dark blue and his skin varied from blue-grey to silver white but no rich or vivid tints appeared. Got his eyes. Their optic nerves have a plicated or folded structure like a ribband doubled longitudinally. . . .

In between all this he read books in a variety of languages and commented furiously on their contents. There was also a considerable social life on board, for among the passengers were Sir Benjamin and Lady D'Urban, going to the Cape to take up the colonial governorship. (A stern governor, he precipitated the Great Trek by which most of the Dutch Boers left the Cape for a British-free life in the interior, was involved in bitter Kaffir Wars on the frontier, and eventually had to be hastily recalled by Whitehall in 1837. His name, however, continues as that of the principal city in the province of Natal.)

1833 gave way to 1834 and they were still at sea, Sir John beginning to grumble at the tedium of it all.

>Wednesday, January 1, 1834
>Commenced the New Year in Lat. 29° South Longitude 11° West on board the Mount Stuart Elphinstone, expecting to arrive at the Cape of Good Hope in 10 days or a fortnight (which God grant!). . . .
>    Baby teething & Mamma has contracted a habit of beating me at chess.—Begin to be tired of keeping a Meteorological Register & wish for sight of land.—Last night at 12 PM being the exit of the old year two disguised persons perambulated the ship banging a large bell—the ladies & gent$^n$ sang Life let us cherish &c—and this morning one of the Cuddy servants was found tied in his hammock, with his face painted black, and quite unconscious of his altered hue.

By January 3 Herschel's grumblings had increased. M was sick again, and Nanson the nursemaid too; there were *no events* on board, and after a spell of "rifle Practice with Capt. Jones, at bottles floated astern & tied by long lines," Herschel was reduced to conducting experiments on the melting point of "cocoa-nut" oil.

As they began to approach Africa's southern tip they encountered that phenomenon bitterly commented on by many another ship's passenger: the Cape rollers. These come from no storm, but are the product of trade winds blowing across the great open reaches of the southern ocean. The huge swells take on a cadence that it seems impossible to accustom oneself to, and ships heave and pitch in the most ghastly manner. I have myself lain clinging desperately to a bunk, night upon night, listening with ever-diminishing care to suitcases being slammed from bulkhead to bulkhead across the floor of the cabin.

The little 611-ton ship undergoing this reduced Maggie to a state of total despair, but most of the attention went to Mrs Nanson, who was in even more desperate straits:

Saturday, January 11, 1834
Towards Evening Nanson who had been regularly mending in health since her last attack, and who, yesterday & today had sate up some time in her armed chair was reported delirious (NB She had talked very loosely & absurdly for 2 days before, but without suspicion of anything more than ordinary weakness of mind. Mr. M^cHardy & D^r Mackintosh declared her delirium a very dangerous indication of increasing weakness and prognosticated a fatal termination. Got all on the alert to sit up with her & keep up the restorative system during the night.

The prognostications soon having proved false, attention switched to preparing the ship for a spick and span entrance to Cape Town, as befitted the vessel carrying the new governor.

Much Painting & Tarring the Ship to make a shew at the Cape, going on—also the gun's ordered to be "sealed" & Cartridges prepared.— NB. Major D. related case of firing a salute from old guns, where from mere neglect of "Stopping the vent" while springing, at the re-loading 2 men were blown limb from limb, & the wife of one was presently seen picking up the scattered legs & arms of her husband (Major D an eye-witness) Also Capt^n R^n had a mans arm blown off in the self same act. Surely this must be a bad system of going to work.

Finally, the long-awaited day arrived:

Wednesday, January 15, 1834
At early dawn this morning James knocked on our cabin door and called out Land! Rose & hurried on Deck whence the whole range of the Mountains of the Cape from Table Bay to the Cape of Good Hope was distinctly seen, as a thin, blue, but clearly defined vapour. The Lions Head was seen as an Island the base being below the horizon. Called up Marg^t who also came on Deck just before Sunrise. It was a truly magnificent Scene. . . . The shore still nearing, grew bolder & more rugged & broken. "The wild Pomp of Mountain Majesty" developed itself & certainly nothing finer than this approach of South Africa can be conceived. . . .

Thursday, January 16, 1834
At daybreak weighed anchor, and got the ship nearer the town into good anchorage about a mile from the jetty. . . .
The situation most remarkable, hemmed in on the sides by steep

promontories & backed by the Mural precipice of the Table Mountain which rises sharp & sudden behind it. We had hardly cast
anchor when a boat came on board with Col¹ Wade to welcome the
Governor and (what was more interesting to us) Marg^ts brother D^r
Duncan Stewart to welcome ourselves. Under his guidance we took
Boats & left the ship & (the Lord be thanked) set foot on African
ground about 10 AM. . . .

Only eleven days had passed since the Maclears had similarly gone ashore
for their greeting by Meadows.

Of course it took some time for the Herschel entourage to settle down.
There were fifteen boatloads of baggage to be unloaded, and clearly a
rather large house would be needed for accommodation.

At first they stayed in a boarding house ("where we were ushered into
*immense* apartments, each large enough to hold an English house—a
small one—& made extremely comfortable") and later in the home of a
Mr Borcherds, the highlight of which was the occasion when Lady Herschel, sitting in the living-room, found a snake coiling itself around her
ankle. Presumably the long-suffering Henderson would have accepted
this with greater equanimity than did she.

But inbetween the unloading and scurrying around looking for somewhere more permanent to live, the Herschels had time to take in something of their new surroundings. By and large they were entranced by the
beauty of the scenery and the multitude of glorious flowers they found
around them. Even the Englishman's solemn duty was not neglected, and
within three days of arriving they attended church, Herschel taking a
rather dry view of proceedings:

> Sunday, January 19, 1834
> After Breakfast attended Divine Service in the Principal English
> Church in this Town.—There are few & small pews the great area is
> occupied by Chairs & benches, in which all sit indiscriminately. This
> is right—In God's sight there should be no exclusive aristocratic
> distinctions—Every one should forget Differences & remember only
> that he is a worm among worms. The New Governor (Sir B.
> D'Urban) was there, with no state. [The historical record does not
> suggest that Sir Benjamin *ever* viewed himself as a worm among
> worms.]
>   The organ Loud & harsh. . . .—A long loud & swift sermon from
> D^r Hough in condemnation of reason and exaltation of faith fol-
> lowed, but he preached so quick it was difficult to *follow him* & I

could only perceive him to be a clever man using words in very terse
& compact combinations as if he *had* a meaning & wanted his hearers
to perceive it.


Before long the Herschels had found themselves a suitable house that
they would later buy for the substantial sum of £3000. It was called
Feldhausen, and was located some six miles out of the town under the
eastern escarpment of Table Mountain, not far from the Constantia
Valley. Sir John described it to his aunt as a "perfect paradise, in the most
rich and magnificent mountain scenery, and sheltered from all winds *even*
the fierce South Easter, by Thick surrounding woods." The Royal Observatory was about four miles away, which was convenient for the in-
cessant traffic that would soon be plying between it and Feldhausen. Sir
John, however, would not be using the observatory's telescopes: he had
brought his own, which were now erected in the grounds of Feldhausen.
It was a spacious estate, and included a small cottage that Herschel used
as a workshop for polishing his mirrors. One room of the cottage became
a laboratory for the inevitable chemical experiments.

Feldhausen would be the Herschel home for the next four years. They
would be tumultuous years, filled with much laughter and tears not only
on the family level (three more of the Herschel children would be born
during this period), but on a larger social level as well. So eminent a pair
as the Herschels, already close friends of the Governor, were very soon
lionized by the small British population at the Cape, and John became
president of all sorts of local literary and scientific societies, including the
Committee for the Exploration of Central Africa, which involved some
interesting planning. They were years which saw much upheaval in the
colony: the British insistence on freeing the slaves, which aggravated an
already tense situation with the Boers; the beginnings of the Great Trek;
the difficult frontier wars which at one stage brought Cape Town under
martial law; the internal political problems involving D'Urban. Herschel
could have escaped little of it, for the diaries show an almost endless
stream of important visitors to Feldhausen.

One major achievement of Herschel on this larger scene was the re-
organizing of the Colony's educational system. The scheme he introduced
lasted into the twentieth century.

All of this, of course, was subsidiary to Herschel's own astronomical
work, which he carried out with great diligence and success. This pro-

The twenty-foot reflector erected by Herschel at "Feldhausen." Delineation by Sir John; lithograph by G. H. Ford

ceeded with few contretemps, although one of these accompanied Herschel's search for Halley's Comet, eagerly expected in 1835. Herschel, furiously sweeping the skies with his telescope, had great difficulty locating it, and when he belatedly did so and made a triumphant pronouncement of the fact to his family, he was taken aback to learn that his assistant Stone had casually seen it with the naked eye several nights before without bothering to report it. ("I never was more inclined to give a man hard words or even a hard knock. . . .")

There was no stopping at astronomy, naturally, and there are endless reports on experiments and observations in just about everything from geology to meteorology to botany to zoology to chemistry. . . . Most of what happened was reported with an ebullience that reveals Herschel's innate boyish enthusiasms. There was the glorious afternoon when Captain Wauchope—fleet captain to the Admiral at Simonstown—came by, and he and Herschel had a wonderful time making explosives and blowing up old tree stumps in the Feldhausen grounds: "[We] had a touch at the huge oak stump in our cross avenue . . . with an ounce vial ⅔ full of a mixture of 4 parts gunpowder & 1 fulminating mercury. It was neatly bisected with an immense bounce & the pieces flew to a good distance." With a mixture like that they were lucky not to have been neatly bisected themselves.

And, of course, there was life with the Maclears. The diaries repeatedly announce that Lady Herschel is spending a couple of days over at the observatory with Mrs Maclear, or that the latter has stopped over at Feldhausen with her brood. One entry in particular is of interest:

> Thursday, February 26, 1835
> Maclear came to make a sweep . . .—At Supper he regaled us with a rich account of his being challenged by a French officer de Sa Majesté—Officier de Marine, for uncourteous conduct (which by his own account appears to have been the fact*
>
> *The names of Truter & Bestandig figured in this relation Arcades ambo [Byron's twist of Virgil's phrase, meaning "blackguards both"]

From which it appears that while Maclear must have got over his ordeal with some sang-froid, his behaviour towards Commander Bosse had been a good deal less civil than the account in his own journal makes out.

As though Herschel did not have enough work of his own to cope with,

*Measurement of the base line for Maclear's Geodetic Survey on the Grand Parade, Cape Town, December 1837. Tentative attribution of this watercolour is to Charles Piazzi Smyth, Maclear's chief assistant at the Observatory until 1845. (By courtesy of the Cape Archives Depot—AG 5747)*

1. Mr Maclear.   2. Major Michell.   3. Lt Williams.   4. Mr Hertzog.   5. Gibbs squarer.   6. Kirke squarer.   7. Policemen, keeping the ground.
8. Cape Spectators.   9. Indian Spectators   10. Members of an Extra-provisional-deputy-superintendent-opposition-sub-committee of the
Right Venerable the Synod of the Ecclesiastical Courts of the Dutch Reformed Church, Cape Town, Cape of Good Hope.
9 etc.

he enthusiastically entered into all of Maclear's projects at the observatory. One visited the other at least every few days, and in between there was a continual exchange of notes and letters, sometimes several in a day. These usually began "My dear Sir," as though the two were not in fact intimate friends.

One of Maclear's tasks was to establish the beginnings of a geodetic survey of the region. The first step was to measure a baseline—a distance of a mile or so—with a precision of a small fraction of an inch. This, of course, was no easy task, involving the laying, bit by bit, of a series of surveyor's rods on trestles, measuring the gaps between the rods, and so forth. Maclear and his assistants plugged away laboriously, while the excitable Herschel would come rushing over frequently to offer advice and suggest new methods of getting the work done. His presence could not always be relied upon:

> My dear Sir,
>      I feel so thoroughly the need of a day's rest after the Met Obsn$^s$ that I must beg you to excuse my not keeping my app$^m$ tomorrow (Thursday) but will . . . come on Saturday (the other being Good Friday)—yesterday's work was too much for me the Therm the whole day being at & above 90 and being exposed for some hours out of doors has produced a degree of over excitement which makes me fear another fit of sinkings something of which I now begin to feel.
>      I send specimens of lines engraved on mica with a steel needle [as a means of measuring the gaps between rods]. . . .
>      Yours truly,
>      JFWH

The work was slow and did not proceed smoothly, a severe set-back coming when the wind blew over one of the rods, which meant that the entire measurement had to be started again from scratch. Sir John hastened to send lugubrious sympathies:

> My dear Sir,
>      I am sorry for the accident but I confess I was at no time sanguine as to the completion of the base without some interruption of the kind. The rods are top heavy and in case of any new operation commenced here with such instruments I should urge on the attention of those concerned a thing which if you recollect I did at one time mention as no imprudent precaution. . . .

My horse yesterday while arranging some matter about his head gear took his stand on my left foot and like an obstinate beast would not get off when I told him—till I was ready to roar out. This has lamed me a little but no bones are broken,
  Yours very truly,
  JFWH
P.S.  The first 700 feet of the Irish base were obliged to be re-measured by an accident similar to yours.

Following hard on his lameness another fit of the sinkings apparently overcame him, for he continued in another note of probably the same day:

I feel today so great an increase of certain very unpleasant sensations which were creeping over me the whole of yesterday, that I hold it very doubtful whether I shall be in a condition to come over tomorrow or at all events for more than an hour or two in the warm part of the day—indeed I begin to perceive that another entire day's duty in your swamp will go nigh to lay me up for the winter, and I cannot help feeling that there are many days work before that measurement will be completed—to say nothing of some unlucky jog which may render it necessary to remeasure the whole.

One wonders what Maclear's private thoughts were on receiving some of these missives. He himself was no stranger to sickness either, as we see from his report of events after a hard night of astronomical observation:

Thursday [April] 10, [1834].  Remarked a peculiar sensation in my mouth on the right side—the feeling of a crumb of bread or something in it. When I blew my nose my cheek filled out and became inflated, while the left preserved the usual muscular resistance so that to do this effectually I am obliged to push in my cheek with my hand. There is no loss of sensation.

10 o'clock.  Took a dose of Epsom salts. On exam$^g$ my face in the glass I find that on laughing the integuments are drawn to the left side and in masticating food the space on the right side between the gums and cheek is filled but I cannot without the pressure of my hand empty it. I cannot effectually close my right eye. For some days I have been working hard with the mural circle. My left eye became tired and painful & 2 days since was rather inflamed, with a pain shooting into my head.

5 a.m.  the Epsom salts operating.

Friday 11.   My face is not better. When I drink or laugh the integuments approach the left side. My right eye is watery and I fancy that the gums of the upper jaw on the same side are painful. Query is the durameter covering the facial branch of the 5th pair of nerves inflamed? Breakfast Tea and Toast.

Saturday 12.   Not finding myself better I sent for Dr Murray. He cupped me and put a blister to the nape of the neck and one behind each [?ear] rubbed the face with [?chile pad]. Took an emetic and applied 12 leeches to the temples. Afterwards nauseating doses of Tartar Emetic and Donen's powder at bed time.

Sunday 13.   Dr Murray called. Took a dose of Lennox salts. Sir John Herschel called to see me. . . .

Monday 14.   My face is much the same. Put another blister over my left cheek which with one on my head yesterday makes five blisters.

Tuesday 15.   Dr Murray called. My face much the same. My left cheek is swelled. Rec^d a consolatory letter from Sir J. Herschel.

This consolatory letter was in one of Sir John's more jocular styles:

I am sorry to find your bulletin speak of no improvement but *"courage mon ami"*—if the worst comes to the worst, a twisted visage may be put up with. In the case of a bachelor on his preferment, it might be a serious thing, but once family taken "for better for worse" and estimated and loved for qualities more than skin deep a man may keep the laugh on his side though it may be only on *one* side.

Adieu.

The crisis lasted until the end of the month, by which time Maclear had survived both his illness and its treatment and was sufficiently recovered to do all but whistle.

Nowhere is the contrast between the dashing and feverishly excitable Herschel and the more stolid down-to-earth Maclear better shown than in a letter concerning Maclear's having to set up a tide gauge.

Herschel as usual was full of novel ideas and had promptly devised a semi-automatic gauge that would ring a bell when the tide was at a certain point. At this point a gentleman whom Herschel termed a "tide-waiter," and who was under orders to "sit still in his guard house and look at his watch," would rise and "take his cloak, lanthorn, key and

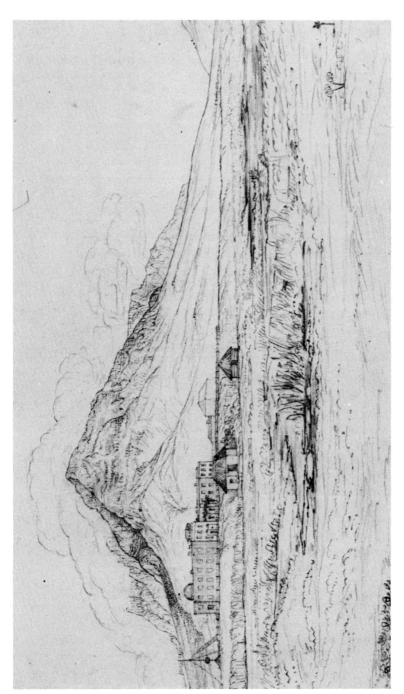

*Camera lucida sketch by Herschel of the Royal Observatory from across the Salt River swamp, dated June 2, 1837 (by permission of the South African Library, Cape Town)*

pencil and go to the gage" to record the time of the event. Unfortunately there was some hitch at one stage, which Sir John reported in one of his most excitable letters:

> My dear Sir,
> I have made so dreadful a *mess* of the Tide Obs$^n$ I took on *Monday* as well of the Meteorological Ob$^{svg}$ by using Daniel's watch—as defies all my power to decipher—i.e. in respect of time. It is something so compounded of the ludicrous & the melancholy as to be an epitome of the great tragi-comedy of human life. In truth I am ashamed to shew it you—yet as a matter of curiosity and philosophical enquiry it may be worth while. I shall therefore ride over to the Observatory between High and Low water tomorrow and "make a clean bosom" of the whole affair.
> Tomorrow (Friday) the stroke of 8 will find me on the jetty. If you will be up, I think I shall take the Observatory in my way and pick up a time I can depend on. My blunders (if blunders they were —i.e. if the evil one had not his finger in the pye) went to whole hours—half hours and quarters. It is something enormous—incredible and to me utterly incomprehensible,
> Yours in haste,
> J.F.W. Herschel
> P.S.   On second thoughts I will not come via the Observatory. There will be too little time and I shall miss the Observations altogether most likely by so doing.

Whether Maclear ever found out what it was all about isn't known, for on the back of the letter there is a pencilled note in his handwriting laconically noting "A fog. Missed his way and found himself at Tyger B." Tigerberg is some five miles away in an entirely different direction.

While Herschel and Maclear were working away at all these multifarious projects, there was being perpetrated half way round the world one of the most extraordinary and famous scientific hoaxes ever devised. It has always since been known as The Great Moon Hoax, and it had as its central figure Sir John Herschel.

The hoax was the work of an obscure journalist, Richard Adams Locke, who was a reporter on a New York newspaper, *The Sun,* but who was British in origin—in fact, a distant descendant of the English philosopher John Locke. Little is known of Richard Locke today, but he was well

enough known in his own time for Edgar Allen Poe to call him a genius.
If the hoax is any evidence, Poe may well have been right.

The whole thing was a parody of Herschel's expedition to South Africa.

According to Locke he had gone there with the most incredible telescope in order to study the moon. Since the telescope was capable of
studying even the insect life on the moon, the way was open for the most
fantastic and chimerical revelations about the earth's satellite.

These revelations appeared as a series of articles in *The Sun,* spanning
less than a week (although totalling some 17,000 words), and seem to
have had almost everyone in the United States agog. Indeed, one of the
fascinating things about the hoax is the degree to which people believed the most ridiculous and absurd of its descriptions. Edgar Allen Poe
later recalled that "Not one person in ten discredited it, and (strangest
point of all!) the doubters were chiefly those who doubted without being
able to say why—the ignorant—those uninformed in astronomy—people
who *would* not believe, because the thing was so novel, so entirely out of
the usual way. A grave Professor of Mathematics in a Virginia college told
me, seriously, that he had no *doubt* of the truth of the whole affair!"

Another intriguing aspect of the hoax was that it contained just enough
in the way of technical material (optics of the telescope and so forth) to
sound plausible, initially at least, to the non-scientists. True, anyone with
astronomical training would have been able to see through it immediately,
but it is something of a mystery how Locke, who had had no scientific
training, was able to insert just the right amount of half-truth. This later
led to the speculation that he had had the help of a visiting French astronomer, but Locke never revealed the truth or otherwise of this. He also
never revealed just why he had perpetrated the hoax; one story claims
he did it as a free-lancer for $150, another that he saw it as a means of
raising his paper's circulation. It does seem, though, that his editor was
as much taken in by the hoax as any of the paper's readers.

The opening instalment appeared in *The Sun* on the morning of Tuesday, August 25, 1835. It purported to be reprinted from a Supplement of
*The Edinburgh Journal of Science,* the original supposedly having been
written by a Dr Grant, who, it was claimed, had been an old friend of
William Herschel's, and who was now accompanying John Herschel on
this great expedition to the South.

The opening paragraphs left the reader in no doubt of the magnitude
of things to come:

> In this unusual addition to our Journal we have the happiness of making known to the British public, and thence to the whole civilized world, recent discoveries in Astronomy which will build an imperishable monument to the age in which we live, and confer upon the present generation of the human race a proud distinction through all future time. . . .
>
> To render our enthusiasm intelligible, we will state at once, that by means of a telescope of vast dimensions and an entirely new principle, the younger Herschel, at his observatory in the Southern Hemisphere, has already made the most extraordinary discoveries. . . .

The telescope, naturally, was of prime importance. Should it be a mirror telescope or a lens telescope? Sir John was an expert on the metallic mirrors of the day, and

> had watched their growing brightness under the hands of the artificer with more anxious hope than ever lover watched the eye of his mistress. . . . He had the satisfaction to know that if he could leap astride a cannon ball, and travel upon its wings of fury for the respectable period of several millions of years, he would not obtain a more enlarged view of the distant stars than he could now possess. . . .

Nevertheless, when it came down to studying the insects on the moon, even this would not suffice. Sir John concluded it would have to be a lens telescope, the lens to be 24′ in diameter (more than seven times larger than any lens ever built in history) and weighing 14,826 lbs. Most lenses have to have several components of differing kinds of glass in order to overcome colour aberrations, but Sir John in a moment of inspiration saw that simply melting all the glasses together "would as completely triumph over all impediments."

But even so he realized that the image formed by this stupendous lens would still require special processing. The problem being beyond even Sir John's inspiration, he travelled up to Edinburgh to consult with Sir David Brewster (who *was* an optical expert in real life).

> The conversation became directed to that all-invincible enemy, the paucity of light in powerful magnifiers. After a few moments' silent thought, Sir John diffidently inquired whether it would not be possible to effect *a transfusion of artificial light through the focal object of vision!* Sir David sprung from his chair in an ecstacy of conviction, and leaping half-way to the ceiling, exclaimed, "Thou art the man!"

Yes, indeed, they would illuminate the image with a hydro-oxygen micro-scope, which would then cast the image on a large canvas screen for all to see as a kind of instantaneous movie.

A working model must be built.

The co-operative philosophers decided that a medium of the purest plate glass (which it is said they obtained, by consent, be it observed, from the shop window of Mons. Desanges, the jeweller to his ex-majesty Charles X., in High Street) was the most eligible they could discover. It answered perfectly. . . .

The real thing, however, would require some financing ("Money, the wings of science as the sinews of war, seemed the only requisite. . . .") But, "fully sanctioned by the high optical authority of Sir David Brewster", Sir John rushed off to lay his plans before the Royal Society. That august body was quickly overcome with fervour. The project

was immediately and enthusiastically approved by the committee chosen to investigate it, and the chairman, who was the Royal President, subscribed his name for a contribution of £10,000, with a promise that he would zealously submit the proposed instrument as a fit object for the privy purse. He did so without delay and His Majesty, on being informed that the estimated expense was £70,-000, naively inquired if the costly instrument would conduce to any improvement in navigation? On being informed that it undoubtedly would, the sailor King promised a *carte blanche* for the amount which might be required.

It turned out that the invaluable Dr Grant had a brother who could cast the lens, and in due course this appeared, "immaculately perfect," after a mere week of cooling.

Now Locke had to find some reason for packing Herschel and company off to South Africa, for he could have studied the moon just as well from the northern hemisphere. The reason for going south, announced Locke, was that Herschel wanted also to observe a transit of Mercury (analogous to a transit of Venus), which would unfortunately occur when it was night in the northern hemisphere but daylight in the southern! Locke managed to confuse even himself at times.

No one seems to have worried about this peculiar reason, however, and the intrepid observers were supposed to have set off from London

on September 4, 1834, there being Herschel and Grant as well as a "Lieutenant Drummond of the Royal Engineers, F.R.A.S., and a large party of
the best English mechanics."

Arrived at the Cape after "an expeditious and agreeable passage", the
company immediately cast about for a suitable site for their observatory.
Sir John mysteriously selected a spot some thirty-five miles north-east of
Cape Town, and the ascent to the plains there was accomplished "by
means of two relief teams of oxen, of eighteen each, in about four days;
and, aided by several companies of Dutch boors, [Herschel] proceeded
at once to the erection of his gigantic fabric."

For openers Sir John decided to get in a shot at Halley's Comet, which
of course would not be visible in ordinary telescopes for many months yet.
Here he seems to have overstepped himself somewhat. Not that he failed
to sight the comet, of course; but he immediately notified others of the
sighting, when his "royal patrons [had] enjoined a masonic taciturnity
upon him and his friends until he should have officially communicated
the results of his great experiment." The great telescope was supposed to
be a secret until the moon results were in. Sir John, however, wrote

to the astronomer-royal of Vienna, to inform him that the portentous
comet predicted for the year 1835, which was to approach so near
this trembling globe that we might hear the roaring of its fires, had
turned upon another scent, and would not even shake a hair of its
tail upon our hunting-grounds. At a loss to conceive by what extra
authority he had made so bold a declaration, the men of science in
Europe who were not acquainted with his secret, regarded his "post-
ponement", as his discovery was termed, with incredulous con-
tumely, and continued to terrorize upon the strength of former predictions.

That was all very well, but *The Sun*'s readers, already grown after the
first instalment from 8,000 to 12,000 a day, were anxious to hear about
the moon. Sir John, though, hesitated. "He was about to crown himself
with a diadem of knowledge which would give him a conscious pre-
eminence above every individual of his species. . . . He paused ere he
broke the seal of the casket which contained it." Fortunately courage and
Dr Grant prevailed, though, and

it was about half-past nine o'clock on the evening of January 10,
the moon having then advanced within four days of her mean libra-

tion, that the astronomer adjusted his instruments for the inspection
of her eastern limb. The whole immense power of his telescope was
applied. . . . The field of view was covered throughout its entire area
with a beautifully distinct, and even vivid representation of basaltic
rock. Its color was a greenish brown. . . . This precipitous shelf was
profusely covered with a dark red flower, "precisely similar" says Dr
Grant, "to the Papaver Rhoeas, or rose-poppy of our sublunary
cornfield; and this was the first organic production of nature, in a
foreign world, ever revealed to the eyes of men."

The basaltic rocks continued to pass over the inclined canvas plane
. . . when a verdant declivity of great beauty appeared. They were
delighted to perceive that novelty, a lunar forest. "The trees", says
Dr Grant, "for a period of ten minutes, were of one unvaried kind,
and unlike any I have seen, except the largest kind of yews in the
English churchyards, which they in some respects resemble".

But trees and flowers were of only limited interest, and Locke began
to introduce something more exciting. He thought of the lunar *maria*.
These dark areas on the moon's face we know to be plains of congealed
lava, but the ancients, not knowing one way or the other, had called them
seas (for which *maria* is the Latin word). Locke was sure his readers
would prefer them to be real seas. The tireless Dr Grant again took up
the saga:

We perceived that we had been insensibly descending, as it were,
a mountainous district of a highly diversified and romantic character,
and that we were on the verge of a lake, or inland sea. . . .

We again slid in our magic lenses to survey the shores of the
Mare Nubium. Fairer shores never angels coasted on a tour of
pleasure. A beach of brilliant white sand, girt with wild castellated
rocks, apparently of green marble . . . and feathered and festooned
at the summit with the clustering foliage of unknown trees, moved
along our [screen] until we were speechless with admiration. The
water, wherever we obtained a view of it, was nearly as blue as that
of the deep ocean, and broke in large white billows upon the strand.
The action of very high tides was quite manifest upon the face of the
cliffs for more than a hundred miles.

Jutting out from the water, the intrepid observers noted, was

a lofty chain of obelisk-shaped, or very slender pyramids, standing
in irregular groups, each composed of about thirty or forty spires.

Dr Herschel shrewdly pronounced them quartz formations, of probably the wine-colored amethyst species. On introducing a lens, his conjecture was fully confirmed; they were monstrous amethysts, of a diluted claret color, glowing in the intensest light of the sun! They varied in height from sixty to ninety feet. . . .

Further along they came on a three-mile island of pure sapphire, and "we frequently saw long lines of yellow metal hanging from the crevices of the horizontal strata in wild net-work, or straight pendant branches. We of course concluded that this was virgin gold. . . ."

*The Sun*'s readers were in an ecstacy rivalling that of Sir David Brewster. Circulation leapt to 19,360 copies, making *The Sun* temporarily the world's largest newspaper, while crowds besieged the paper's offices, clamouring for copies as fast as the steam presses could provide them. While waiting, the mob was assured by a highly respectable-looking gentleman among them that he had actually seen the telescope being shipped in England, and that the whole thing was undoubtedly true.

The burning question in everyone's mind, however, was what about animal life on the moon? They didn't have long to wait:

Small collections of trees were scattered about the whole of the luxuriant area; and here our magnifiers blest our panting hopes with specimens of conscious existence. In the shade of the woods on the south-eastern side, we beheld continuous herds of brown quadrupeds, having all the external characteristics of the bison. It had, however, one widely distinctive feature, which we afterwards found common to nearly every lunar quadruped we have discovered; namely, a remarkably fleshy appendage over the eyes, crossing the whole breadth of the forehead and united to the ears. This hairy veil . . . was lifted and lowered by means of the ears. It immediately occurred to the acute mind of Dr Herschel, that this was a providential contrivance to protect the eyes of the animal from the great extremes of light and darkness to which all the inhabitants of our side of the moon are periodically subjected.

Locke must have thoroughly enjoyed himself writing about the lunar animals. There were unicorns "of a bluish lead color; it was gregarious, and chiefly abounded on the acclivitous glades of the woods." In a delightful valley containing a large branching river were to be found numerous kinds of water-bird engaged in "pisciverous experiments," although unaccountably the lunar fish escaped Herschel's gaze. At times even Locke's imagination ran away with him, as in the case of the bi-ped beaver:

> The last resembles the beaver of the earth in every respect than in its destitution of a tail, and in its invariable habit of walking upon only two feet. It carries its young in its arms like a human being, and moves with an easy gliding motion. Its huts are constructed better and higher than those of many tribes of human savages, and from the appearance of smoke in nearly all of them, there is no doubt of its being acquainted with the use of fire.

Dared one hope for anything more human-like than a mere beaver? Locke was not one to deny his readers:

> We were thrilled with astonishment to perceive four successive flocks of large winged creatures, wholly unlike any kind of birds, descend with a slow even motion from the cliffs on the western side, and alight upon the plain. They were first noticed by Dr Herschel, who exclaimed, "Now gentlemen, my theories against your proofs, which you have often found a pretty even bet, we have here something worth looking at: I was confident that if ever we found beings in human shape, it would be in this longitude. . . ." Certainly they were like human beings, for their wings had now disappeared, and their attitude in walking was both erect and dignified. . . . They averaged four feet in height, with short and glossy copper-colored hair, and had wings composed of a thin membrane, without hair, lying snugly upon their backs, from the tops of the shoulders to the calves of the legs. The face, which was of a yellowish flesh color, was a slight improvement upon that of the large orang outang. . . . Lieut. Drummond said they would look as well on a parade ground as some of the old cockney militia!
>
> Whilst passing across the canvas, and whenever we afterwards saw them, these creatures were evidently engaged in conversation; their gesticulation, more particularly the varied action of their hands and arms, appeared impassioned and emphatic. We hence inferred that they were rational beings. . . .
>
> We scientifically denominated them the Vespertilio-homo, or man-bat; and they are doubtless innocent and happy creatures, notwithstanding that some of their amusements would but ill comport with our terrestrial notions of decorum.

Clearly this had to be the highpoint of the series, and Locke now had to find a way of winding it all down. Already moonset was tending to

"Lunar Animals." A lithograph copyrighted by Benjamin Day, owner of the New York Sun, which accompanied the paper's Moon Hoax series in 1835 (by courtesy of the Library of Congress)

come rather early on most nights, although on such occasions Herschel and Grant, their minds "being actually fatigued with the excitement of the high enjoyments we had partaken," nevertheless would thoughtfully "call in the assistants at the lens, and reward their vigilant attention with congratulatory bumpers of the best 'East India Particular,' " a courtesy rather regrettably neglected by modern astronomers. The congratulatory bumpers went down so well on one such occasion that the assistants carelessly left the giant lens exposed all night. They were called the next morning by some "domesticated Hottentots" to find that the sun, shining through the lens, had burnt down a clump of trees and "vitrified to blue glass" the plaster of the observatory's walls. In the end Locke was reduced to falling back on sinister injunctions from Dr Grant that certain "highly curious passages" were not to be published, adding that "it is true, the omitted paragraphs contain facts which would be wholly incredible to readers."

The denouement came rather more swiftly than Locke had intended, however. *The Sun* had repeatedly hinted that there were all sorts of technical and mathematical details that they were omitting from these articles for the general public, and this brought two professors scurrying down from Yale to see the technical reports for themselves. The alarmed Locke sent them off to another New York address, and then himself doubled through backstreets to ensure they would again be sent to a third address, and so on. It didn't take the professors long to realize what was happening, nor Locke to realize that the game was up. A rival New York newspaper, the *Journal of Commerce,* swallowing its pride in favour of its circulation, had asked permission to reprint the entire series, and now to it Locke suddenly confessed that the whole thing was a hoax. The *Journal* managed to bolster both pride and circulation by triumphantly announcing this fact to a startled public.

What were the repercussions of The Great Moon Hoax, and in particular, what did the real Sir John Herschel have to say about it when he heard? The answer, on both counts appears to be: surprisingly little.

The American public seems to have been more amused than upset by the whole affair, apparently taking it as a joke against the astronomical Establishment rather than against themselves. Richard Locke, however, did very soon part company with *The Sun,* to pursue his journalistic career elsewhere.

The hoax was widely reprinted in pamphlet form and soon appeared across the Atlantic. But with news travelling so slowly then, and since the entire episode had been over in hardly a week, it was known to be a hoax by the time the Europeans read it. Accordingly, reaction was mild. The British, even those in Edinburgh, managed to keep a pretty stiff upper lip about the whole affair, perhaps allowing themselves a shake of the head over the deplorable American press. The more volatile French, however, seem not to have been at all amused, the French Academy— particularly its president François Arago—finding itself outraged over this besmirching of John Herschel's good name.

It would hardly have been surprising if Herschel himself had felt a considerable degree of outrage. It seems he learnt of the hoax through a Mr Caleb Weeks, an American proprietor of a menagerie who arrived in Cape Town, a copy of the hoax in hand, to buy African animals. Story has it that Herschel was by coincidence in Weeks' Cape Town hotel when he heard Weeks asking at the desk where he might find the great Sir John Herschel. On the latter introducing himself, Weeks is supposed to have said, tongue in cheek, that he was so pleased to meet the famous astronomer whose discoveries he had been reading so much about in the New York newspapers. Sir John, of course, was considerably taken aback, saying that he had not yet announced any of his new discoveries, and no doubt privately wondering which among them could possibly have left New York daily readers agog in any case. Weeks then handed Herschel a copy of the hoax pamphlet and left the great man alone to read it over. Sir John was very quickly at Weeks' elbow again, agitated in case the hoax had really originated in Edinburgh, but on Weeks assuring him that it was merely the work of an *American* paper, Sir John supposedly laughed and relaxed and invited Weeks to sit down and tell him all the details.

That at least was Weeks' version of the matter, as retold a number of years later. The usually excitable and loquacious Herschel was surprisingly mum about the whole affair. There is not one word given to it in the diaries, and the only direct Herschel references to it come in two letters written to Aunt Caroline. The first, undated but received by Caroline on October 1, 1836, was written by Maggie. The bulk of the letter is given over to family news, but Maggie goes on to enthuse over a pretty star map that Herschel is working on, and then quite suddenly, without a new paragraph, says

Have you seen a very clever piece of imagination in an American Newspaper, giving an account of Herschel's voyage to the Cape with an Instrument [omitted] feet in length, & of his wonderful lunar discoveries Birds, beasts & fishes of strange shape, landscapes of every colouring, extraordinary scenes of lunar vegetation, & groupes of the reasonable inhabitants of the Moon with wings at their backs all pass in review before his & his companions' astonished gaze—The whole description is so well *clenched* with minute details of workmanship & names of individuals boldly referred to, that the New Yorkists were not to be blamed for actually believing it as they did for forty-eight hours—It is only a great pity that it *is not true,* but if grandsons stride on as grandfathers *have* done, as wonderful things may yet be accomplished.

The other reference to the hoax appears in a letter from John to Caroline on January 10, 1837. It is a lengthy letter, again filled with family news and details of John's work. Only in a postscript is there brief mention of the hoax:

P.S.   I have been pestered from all quarters with that ridiculous hoax about the Moon—in English French Italian & German!!

And that was the victim's last, perhaps only, word on the subject.

The Herschels' four years in South Africa slowly drew to a close. John's basic mission, which was to survey the southern skies for nebulae, star clusters, and so forth, was completed with great success. His results, properly tabulated and published after his return to England, would later be edited together with those of his father into a great New General Catalogue of clusters and nebulae, and to this very day the listings of the so-called NGC are in constant use by astronomers everywhere.

He accomplished much more besides. Apart from all the novelties he introduced into Maclear's work, he gave the first detailed descriptions of the Magellanic Clouds (the Milky Way's nearest companions in space), studied Halley's Comet, calibrated the scale of star magnitudes used by astronomers, and was present to observe and record a unique event in astronomy, the explosion of a strange star called eta Carinae.

Finally the family arranged to sail aboard the *Windsor* in the southern autumn of 1838. Characteristically, there was a tremendous rush to get everything packed in time.

> Sunday, March 11, 1838
> Rose at ¼ before 1 AM and with great exertion completed final packing & all arrangements for leaving Feldhausen. D$^r$ Liesching & M$^r$ Smith (Surgeon of the Windsor) reported Willy fit for going on board, so putting him in Blankets with Mamma & Mrs Hutchinson in the Carriage and the children & Nurses (Macqueen & M$^{rs}$ Humphries Leah & Catherine) into Baron Ludwig's Waggon, packing & dispatching the cart full of Baggage with Stone & M$^{rs}$ Hutchinson's Covered Cart also loaded with James as Driver, mounted "Old Harry" [his horse] and after Brief adieus Galloped away Down the glorious avenue of Feldhausen—never turning a head to look back. . . .

The return journey was much a copy of the one four years before, although Maggie seems to have withstood it much better. This may have been in part because her husband soon turned his attentions to devising an improved hammock arrangement for sleeping:

> Devised a perfect Swing Cot and proceeded to fit ours up on the principle. The rolling & the pitching are both separately & independently corrected. PQ Cross Beam of Cabin cieling. . . .

As usual, Sir John had a go at everything he could think of. The capture of a shark offered opportunity for an orgy of dissection ("Out of it were taken 24 *live* young sharks, which although the umbilical cord was still attached, yet lived and were very active in a pail of sea water. Put three of them in Brandy"), and when he could think of nothing else scientific or literary to do, he engaged in trials of physical strength and endurance among the male passengers on deck—not too successfully.

Returned to England, Herschel's fame steadily increased, although his career as an observing astronomer was over. He never set eye to telescope again. But there was all his South African work to be written up, and as the years went by he became more and more the elder statesman of science, being much in demand for all kinds of important committees and presidencies. The new young Queen Victoria made him a baronet. Later he became Master of the British Mint, where he tried unsuccessfully to introduce a system of decimal coinage into Britain. Herschel's interests never slackened, and into old age he worked incessantly on an enormous range of projects, at home sometimes amusing himself by translating Latin, Greek, Italian and German poetry into English hexameters. It is difficult today to realize the esteem in which he was held, although it has been said of him that he was to the nineteenth-century man-in-the-street what Einstein became in the twentieth century. Certain it is that when he died in 1871 his countrymen saw fit to bury him not merely in Westminster Abbey, but alongside Isaac Newton at that. Yet his very universality was in the end his undoing, for outside astronomy, the tide of history has washed over him with scarcely a ripple.

Feldhausen survives. It is today a girls' school (named, of course, "Herschel") in the heart of the busy Cape Town suburb of Claremont, and not far off stands an obelisk that marks the spot where Herschel's telescope stood. Heavy traffic honks its way southwards over a thoroughfare still known as "Herschel's Walk."

Thomas Maclear spent the rest of his life at the Cape, retiring from the directorship of the Royal Observatory in 1870, and living quietly in a nearby suburb until his death in 1879. He and Herschel never dropped their long friendship, corresponding frequently, and it was Maclear who wrote Herschel's principal obituary in 1871. Maclear's career, as one would expect, never came close to matching that of the dashing Herschel in terms of fame and limelight, although he was said to have been much revered by the colonists in his later years, cheers breaking out at his mere appearance at public meetings.

The famous baseline was finally measured, and from it Maclear started off on a large-scale geodetic survey of the colony which perhaps only someone as dogged as he could have brought to a successful conclusion. The survey was conducted with an enormous instrument for measuring angles called a sector, which had been built for entirely different purposes by an earlier Astronomer Royal, James Bradley, in England many years before. It needed a tent seventeen feet high for its use, and with it Maclear went lumbering around by ox-wagon to remote parts of South Africa, carrying out his survey. There are still geographic features in odd corners of the country which bear names like Sectorberg, no doubt to the complete mystification of the local inhabitants.

If Herschel ever grew tired in the midst of his ceaseless activities, perhaps he gave an envious thought to Maclear's quiet life. For he once wrote, shortly before leaving Africa,

> Whether much good—or much evil is in store for us on our return to Europe is uncertain—but this I know that we have been very happy here—and that our residence at the Cape—come what may will always be to me as Malthus somewhere or other beautifully calls some such happiness—the Sunny Spot in my whole life where my imagination will always love to bask.

# A SCOTSMAN ABROAD

*David Gill in Search of the Solar Parallax*

*David Gill in 1884, from the portrait by Sir George Reid. This appears in* David Gill: Man and Astronomer *by George Forbes, London 1916, and is reproduced by courtesy of John Murray (Publishers) Ltd.*

We came into Sal. The Southern Cross hung slantwise in the black sky, aimed dagger-like at the hot coast of West Africa. It was 4 am, and we breathed in great lungfuls of the tropical salty breeze after the long flight from Hamburg, filing stiffly down the metal steps and making our way across the apron towards the crude little terminal building. The crowd at the bar jostled for the barmen's attention, irritated at there being only warm Portuguese beer to drink. Claiming a bottle and paper cup I turned to find a seat at the crowded little wooden tables, over which the cigarette smoke was already rising in a haze. Too late I saw my seating companion from the plane waving me over enthusiastically to an empty chair beside him.

Middle-aged, portly, his grey suit rumpled from continuous wear, he was returning to Johannesburg after business meetings in Europe. For a while I managed to stall him, and we sat watching the groundcrew clattering around as they refuelled the Jumbo. But he was of that irrepressible breed who remain cheerful and talkative whatever the hour or circumstance, and now, his strong hand pouring the last of his beer into the paper cup, he took up again our previous dinner conversation.

"What I can't see, man," he began, his flat Afrikaans accent elongating the a's, "is how you astronomer blokes can find out things like how far away the sun is. I mean it's millions and millions of miles, no? So how do you do it? Can you tell me just simply, hey man?"

I studied my beer. Despite years of teaching astronomy, my fatigued brain refused to locate the logical starting point for an answer. Instead it began to review the improbability of finding oneself on a forlorn little island in the Atlantic, drinking warm beer at 4 am, and being interrogated by a persistent inquisitor on the subject of the sun's distance. And the association of Atlantic islands and the sun's distance suddenly brought to mind something else. Davie.

That early morning in 1958. The little freighter grinding slowly across the Atlantic, two weeks and five thousand miles out of New York, with the grey dolphins diving carelessly through the waves ahead as though in escort to our first landfall. We passengers clustering eagerly on deck to watch the finger of the dead volcanic cone rise slowly above the horizon. Ascension Island, a small wasteland of lava only half a dozen miles across. As in Davie's day, there was no port, but we sailed close inshore, enough to see the waves pounding the bleak black rock. There came a "whoop-whoop" of greeting from the shore battery, to which we replied with a deafening blast of the foghorn, and then passed on down the brief coast, bound for St Helena.

As we rounded a headland, the deck officer nodded towards a little white beach suddenly revealed among the dark cliffs. "Mars Bay," he said. I don't suppose it had changed much since Davie was there.

Twenty-six North Silver Street was a comfortable but rather ugly little house, and the furniture, which I thought beautiful and David did not think about at all, atrocious. But to us both a very heaven of happiness lay between its four walls, as it always did between every four walls which held us two. . . .

The writer was Mrs. David Gill, or Isobel to give her her own name, although she and her husband more usually referred to one another as Davie and Bella. She was describing the early years of their marriage, when twenty-eight-year-old David was the proprietor of a prosperous watchmaking business in the High Street of Aberdeen, Scotland. It was 1871, and David, competent and happy in his work, was very much a rising businessman, already making £1,500 a year. His wife describes him thus:

He was then a fine young fellow of medium height, slight, with a supple boyish figure, carelessly dressed, quick of movement, with dark brown hair much dishevelled, from a habit which never left him, of constantly passing his fingers through it, and a twist of humour hovered about the mouth and also twinkled in the eyes. In the eyes, however, there was more than humour. There was a compelling power, an unconscious strength which held one, and which showed that, although unconscious of it, he had already found himself. . . . Although probably to many ears his voice was not a melodious one, being loudly pitched with a very pronounced Scottish accent, yet I have no hesitation in saying, it was the compelling quality of his voice, with its extraordinary variety of tone, which expressed his individuality in a way that made the listener, without knowing why, listen to him and remember what he said. There seemed to be no emotion that it could not express. . . .

The lives of David and Isobel were changed for ever on December 12 of 1871. The winter gloom of early darkness had long since set in when David returned to their small house that evening. He and Isobel had

scarcely finished dinner when a ring of the doorbell announced the un-
expected arrival of a young curate, the private chaplain of the Earl of
Crawford. He had brought a letter from his employer's son, Lord Lindsay,
but David set it aside while they chatted amiably for a few minutes. After
the visitor had left, David took up the letter and sat very still reading it.
Isobel watched his face expectantly, until silently, with controlled emotion, he handed the letter to her to read. It announced that Lord Lindsay
was intending to set up a private observatory on the family estate of Dun
Echt, an observatory to rival Greenwich in equipment, and that he very
much hoped David Gill would accept an appointment as the observatory's first director. The salary would be £300 a year. It needed scarcely
a glance between David and Isobel to know the answer, and the fivefold
cut in income was hardly mentioned. David accepted.

David Gill was descended from a long line of Scottish watchmakers,
and his stern and certainly dour father was quite unyielding in his determination that David would continue the family tradition. He was the
third of seven children, but since his two older brothers had died in
infancy, he, as the oldest surviving son, must continue the family business.
When the boy showed modest intellectual talents and an interest in things
academic, his father grudgingly allowed him to enrol in Marischal College of Aberdeen University. But only as a private student, which meant
that he would not be eligible for a degree, and only on the strict understanding that such pursuits would not sway him from his intended career
as a watchmaker.

But the son's taste of the forbidden fruit proved to be the ruination of
his father's plans. For at Aberdeen University David came under the
influence of one of the great intellectual giants of the nineteenth century:
James Clerk Maxwell, already hailed as one of the finest physicists since
Newton. Maxwell, though, must have seemed an unlikely source of inspiration at first meeting. Another of his students said of him: "Judged by
ordinary standards, Maxwell was not a successful lecturer; but there were
some students who could catch a part of his meaning as he thought aloud
at the blackboard." David himself explained:

> Maxwell's lectures were, as a rule, most carefully arranged and writ-
> ten out, and we were allowed to copy them. In lecturing he would
> begin reading his manuscript, but at the end of five minutes or so he
> would stop, remarking, "Perhaps I might explain this," and then he
> would run off after some idea which had just flashed upon his mind,

thinking aloud as he covered the blackboard with figures and sym-
bols, and generally outrunning the comprehension of the best of us.
Then he would return to his manuscript, but by this time the lecture
hour was nearly over and the remainder of the subject was dropped.
. . . Perhaps there were a few experimental illustrations—and they
very often failed. . . .

But those of us who chose to stay behind after the class used to get
a most delightful hour or two, and learn an immense deal that we
never forgot—a great deal that we did not understand at the time,
but that came back to us afterwards—until Mrs Clark Maxwell
arrived, wondering why the professor had not come home to his
dinner, and carried him away *nolens volens.*

Somehow Maxwell's lectures managed to include a few topics on practical astronomy, and with these David found his imagination particularly
fired. By 1863 he had arranged an introduction for a visit to the Royal
Observatory in Edinburgh, and here he turned up one afternoon to be
cordially entertained by the eccentric Piazzi Smyth, then Astronomer
Royal for Scotland. David was entranced by this his first visit to a real
astronomical observatory. With his background in watchmaking he particularly enthused over the observatory's timekeeping instruments, and
questioned Smyth closely on their workings.

Returning to Aberdeen he became determined that the city should
have its own timekeeping service, to be provided by his own observations.
With the help of a sympathetic professor he found that the University
owned the necessary equipment, although it had long since been allowed
to decay into dereliction. The skilled David, however, restored it to proper
working condition, and from then on was an ardent observer of the
heavens.

But his father was not to be denied. At the end of 1863 David was
forced to leave the university and made to get down to the serious work
of watchmaking. His father packed him off to learn the trade, first in
London, and then in Switzerland. Here he at least became fluent in
French, although of a kind referred to by one of his friends as "Gill's
Aberdonian French." Indeed, his broad Scottish accent, which became
even broader with excitement, was a never-ending source of amusement
to many. Later in life he was once haranguing a German astronomer in
English on some controversial point, when, suddenly doubting his companion's comprehension, he broke off to exclaim "You do understand,
don't you?"

"I'm afraid not," replied the other solemnly, "I speak only German and English."

In Aberdeen once more, David settled down to long years as an assistant, later a partner, to his father. Not that they were unhappy years; David was far too innately cheerful and full of life for that. He enjoyed a good social life, looked forward to dances and the company of women, and was not at all averse to after-dinner brandy and cigars. It was discovered that he had considerable talents as a sharpshooter—sufficient to put him on international teams in rifle competitions, although most of this talent was turned to hunting in the Scottish Highlands.

His father, however, does not seem t   .ve approved much of this lifestyle, and a particular conflict       in 1865 when the twenty-two-year-old David announced tha`. he wished to marry sixteen-year-old Isobel Black. Old Mr Gill (he was already seventy-six) was violently opposed, and the confrontation reached such proportions that a family friend reported that "this was the only occasion on which I ever saw young David lose command of hand and eye through the violence of his anger." But his father prevailed, and the young couple had to wait another five years until Gill Senior had finally retired and David had the family business to himself before they were married.

Throughout this time, though, David's enthusiasm for astronomy never diminished. Many of his evenings were spent in extending his previous training in mathematics and physics towards astronomy, and by now he was able to afford a telescope of his own. The turning point came in 1871, when David began a program of photographing the moon, at that time still something of a novelty. It was one of his excellent lunar photographs that came to the attention of Lord Lindsay, and Lindsay, impressed by reports of the young man's abilities and enthusiasm, decided he would make an excellent director of the great new Dun Echt Observatory.

Although Lindsay kept a firm hand on his observatory, and insisted on making almost every decision himself, he did pay close attention to David's recommendations. He even went so far as to send David off on a tour of the major observatories of Europe in order to decide on the best equipment, a tour that proved invaluable to the friendly and outgoing David in making contacts among the astronomical great of his day.

1872, when David moved to Dun Echt, was a critical year in astronomy, and it would decide the course of his entire career. It was the year in which serious preparations first began for the next transit of Venus in

1874, an event to which almost every astronomer was eagerly looking forward in the hopes that it would provide the definitive answer about the distance of the sun. So, right from the start, David became caught up in the problem of the sun's distance, and his fascination with this eventually made him one of world's top authorities on the matter.

Lindsay was as enthusiastic as anyone over the forthcoming transit, and soon made up his mind to mount an expedition of his own for its observation. He also accepted David's suggestion that one of the best instruments for the purpose would be a heliometer. This now obsolete instrument was a small telescope specially adapted to measuring angles with great accuracy. In the hands of David, trained in working with small high-precision instruments, it became the nineteenth century's finest tool for this kind of astronomy.

Where should the expedition go to observe the transit? After consultation with the planners of other expeditions, the site chosen—probably with unconscious irony—was Le Gentil's *bête noire,* Isle de France, although since it had become a British Possession after the Napoleonic Wars it was now known as Mauritius.

Lord Lindsay was pleased with the choice, since it meant that the party could travel out aboard his private 380-ton yacht, the *Venus.* Later, however, there was a small change in plan. Lindsay had discovered that one of the major problems with the eighteenth-century transits had been the uncertainties in the observers' longitudes, and not being a man who dealt in half-measures, he now ordered up no less than fifty of the finest chronometers to overcome any similar problem on his expedition. It was deemed prudent to send the chronometers by the most direct route to Mauritius, which meant going out via Egypt and Aden, whereas the *Venus* would follow the traditional route around the Cape of Good Hope. So it was that David was designated to travel alone with the chronometers, while Lindsay and his friends took a more leisurely approach on the yacht.

The mere job of keeping the fifty chronometers properly wound would have been task enough, let alone ensuring that no harm came to them during the journey. David describes fetching the chronometers from the Royal Observatory at Greenwich, where they had been undergoing rating tests.

> I had no assistant, not even a servant, to accompany me. The chronometers were packed in six well-stuffed inner cases, and these were

suspended by india-rubber bands in strong outer cases, the latter when at sea being separately mounted in large gimbals, swinging between strong wooden uprights fixed in a frame which could be secured to the floor of the cabin. Before my visit to Greenwich I had no misgivings as to the success of the arrangements. But when Sir George Airy [the Astronomer Royal] and his assistants came to wish me good-bye, and when they saw me go off with all the chronometers on the top of two cabs, and no one but myself to look after them, it was evident that they regarded the whole matter as an experiment of very doubtful success, and did not envy the task before me. A first suspicion of these difficulties dawned on me at the time, and they were fully realized before the expedition was over.

In the event, though, all went well, and the morning of August 4, 1874 saw David and his chronometers arriving in Port Louis, Mauritius. There was no sign of Lord Lindsay, so David addressed himself to the question of choosing a site for their operations.

Mauritius, now as then, is a delightful island, and as one sits comfortably at the window of a large jet, sweeping in over the wide white beaches and up across the expanses of sugarcane-clad hillsides, one is struck by its singular beauty. Even so, the ominous clouds which hang frequently over its peaks are an ill omen for the astronomer, and David was lucky in choosing one of the island's better sites. This was the large estate and plantation of Belmont, whose owner, a Monsieur de Chazal, hospitably offered it for the purpose. How M. de Chazal's creole French coped with Gill's Aberdonian French is not recorded. (I was once told with considerable indignation by a Mauritian that when she visited Paris it had been suggested to her that she express herself in English.)

Apart from winding his chronometers, there was little else that David could do astronomically until Lindsay arrived with the other instruments. He was, however, turning over in his mind an entirely new approach to the question of measuring the sun's distance. For the truth was that David did not share the general enthusiasm for the transit of Venus; already he could foresee that the transits would never lead to a truly accurate value. Instead, he was thinking about a method which would involve only one observer making measures with the heliometer on a nearby planet or asteroid through the course of a night. Rather than send several observers to remote parts of the earth in order to provide a baseline for the measurements, why not let the daily rotation of the earth sweep one observer across thousands of miles of space to provide the baseline? Although he certainly intended doing his best at the transit observations, he also hoped that there would be an opportunity while at Mauritius at least to try out this new method. If only Lindsay would arrive with the heliometer.

Meanwhile the sociable David was enjoying himself, as he usually did everywhere. There was the novelty of an invitation to go spear fishing on the coral reefs by night:

> Having accepted an invitation from Rudolph de Chazal to spear fish on the reefs around Amber Island, we set off one evening about five o'clock. The carriage took us to the beach, and the boat to the Island. Here we put on old clothes, and sailed for the reef about a mile out to sea. The flambeaux (great bundles of small pitchy sticks bound together) are lit and we step out upon the reef up to the knees in water. Here we separate into parties of two or three, each party being accompanied by a black fellow carrying a lighted flambeau over his shoulder. It was already quite dark. Each flambeau lighted up clearly a little space around, showing the dark waves breaking white on the reefs, becoming still and green as they pass inside the basin. In the distance each party in its illuminated circle is seen clear and distinct, or a sportsman stopped over a pool with uplifted spear ready to strike.
>
> But this strange effect of light is not so strange as the scene under foot. The reef is like a road, broken up by deep pools and fissures and flooded with water—but what a road! So beautiful! so variegated! Coral every shape and colour, wonderful animals, bunches of seaweed—all the wonders of tropical submarine life—new to me and beautiful.
>
> Now for our proper work, we come to a hole and I see only a wonderful natural aquarium—"See," says Rudolph, but I see nothing. "Ah, he is gone," and I was too late, again and again too late, and then at the next hole I thought I saw a curious long blue stone, it moves, down plunges the spear and a struggling at the end told I had struck. "Keep him down, keep him down to the bottom", cried Rudolph—"now"—and up with the spear came a large blue fish with a bill like a parrot. The Malabari took him off and put him in a bag and we passed on.

Lindsay and his friends had been having a good time of it too. There was much in the way of deck sports on the yacht to keep them amused, and sometimes remembering the seriousness of their mission, there was practice training for everyone, including the deckhands, as auxiliary transit observers.

All went well until they had rounded the Cape and were in the Indian Ocean, when—what would Le Gentil have said?—they ran into alternately heavy headwinds and periods of total calm. It was this that caused the excessive delay in reaching Mauritius, although they were lucky not to have had worse.

Finally, Lindsay in exasperation left the yacht to make what time it could and went on himself by steam cutter. He arrived dishevelled and unannounced in Port Louis on November 1, scarcely a month before the transit. It was evening and Lord Lindsay was much in need of sustenance, but the *patron* of the local hotel was unimpressed by this lanky, red-bearded Englishman, disarrayed, eyes weary behind blue spectacles. A local newspaper reporter gleefully wrote up the exchange:

—J'ai du rostbeef. . . .
—Nô!
—J'ai du plumpudding. . . .
—Nô!!
—J'ai du jambon. . . .
—Nô!!! Donne-moi des sandwitches.

Finally, "le flegme britannique" wore thin, and Lindsay imperiously announced his full title:

—Lord Lindsay! . . . Milord! pardon! Je ne savais pas!

The *Venus* arrived a week or so later, and David and the assistants quickly erected the heliometer and other instruments. It was already past the specific date that would have been most desirable for testing David's new method of finding the sun's distance, and now, after weeks of beautiful weather, there came a good deal of cloud. Nevertheless, he got what observations he could, and then all preparations were turned to the forthcoming transit on December 8. Luck was with them, for the day was clear and calm, and the Lindsay expedition obtained as good measurements as anyone. But David would be enormously pleased when, back in Scotland many months later, he made the necessary calculations from his observations to find that his new method, despite the incomplete and less-than-desirable observations, gave a result as good as that obtained from the transit itself. It showed that one observer under good circumstances could do better than whole armies of observers sent at tremendous expense and labour to remote parts of the world. And the method could be applied at frequent opportunities, rather than having to wait for the very rare transits of Venus.

But for the moment David would have to get the fifty chronometers back to Dun Echt. Lindsay set off on his yacht again, while David and the chronometers retraced their earlier journey. One reason for the time-pieces not going back on the yacht was that David hoped to use them for establishing better longitudes of various islands off East Africa. The sort of difficulties he encountered are described in a letter he wrote to Lady Crawford, Lindsay's mother, from Aden on January 23, 1875.

I was very anxious to determine the longitude of Mahi, the capital of the Seychelles.

I had to land an instrument on arrival and determine time by the sun or stars or whatever I could get,—but because of measles at Bomba we were put in quarantine. I then applied to the Captain to be allowed to land on the Quarantine Island. . . . The little Captain, however, was in such a rage at being put in quarantine that he would not allow me to land, or what was the same thing would not give me a boat to go, and I was in despair when to my great delight I heard the cheerful voice of my friend Captain Wharton—"Hullo Gill, are you there?"—He had been detained on some surveys of reefs on his way to the Seychelles, and hearing we would be put in Quarantine had turned out all his officers during the day to observe equal alti-tudes of the sun for time, and had come off himself to get one of Lord Lindsay's chronometers for comparison with his own.

He had previously obtained permission from the health officer to be allowed to receive two chronometers *"if they were previously dis-infected"*. I applied to the Captain to send the chronometers.— "What! chronometers! send chronometers! Where? Who? What? Have I not told you you cannot go?"

"I don't wish to go—only to send chronometers."

"But you cannot—impossible—quite impossible."

"Will you not assist me?"

"No, I shall not. Why should I?"

"In the cause of science."

"I know nothing of science, only money."

(I must tell you the Captain always when excited takes every astronomer for a Prussian because he has four Prussian astronomers on board and it is almost too much for him.)

"But it is of importance."

"Important or not it is nothing to me. They have put me in quar-antine, and am I to break my quarantine for your sake?"

"Well, Captain Wharton is there and has obtained permission to take two chronometers from the ship."

"Eh—what—obtained permission you say?"

"Yes."

"To land chronometers! They make exceptions for him and they will not allow me to land anything. Very good, I will write to the Governor."

"Yes, I think you are quite right. You should allow me to land the chronometers, and then you will have good cause of complaint."

"Exactly."

So the chronometers having been duly rubbed with vinegar were put in a boat and dropped astern when Captain Wharton received them, and the little Captain retired to his cabin where I saw him for a long time furiously composing letters to the Governor. . . .

It was in Egypt on the way back that circumstances arose which saw a parting of the ways between David and Lindsay. The Khedive of Egypt was most anxious to have a geodetic survey made of the country, and pressed David to stay and become chief of this operation. David wrote to Lindsay (at this stage in Florence, recuperating from his adventures), asking either his permission for a lengthy stay in Egypt, or gently suggesting that perhaps Lindsay would prefer to find another director for Dun Echt. There was more than just the Egyptian operation involved. On David's part there was the realization that he was a man who was going places, was already highly thought of by world-renowned astronomers, and he found it uncongenial to be in a position where his every move required permission from Lindsay. On Lindsay's part, there was considerable pressure from Lady Crawford against David; she preferred to think of the latter as her son's tame astronomer, and didn't like the attention he was attracting when she felt Lindsay should be obtaining more recognition. So Lindsay replied that perhaps the time had come when David should resign his directorship of Dun Echt. Nevertheless, they parted very amicably, and in fact it would be some time before David returned to Dun Echt to wind up his affairs.

It might be said in passing that the Dun Echt Observatory continued under other directors for another quarter-century, until eventually Lindsay, then become the Earl of Crawford, lost interest in it and transferred the instruments and library to Blackford Hill in Edinburgh, where they became a valuable addition to Scotland's Royal Observatory.

In Egypt David was at long last re-united with Isobel, who all this while had been staying at Cannes. Their sojourn in Egypt, though, would not be as long as they expected.

The first order of business was to measure the inevitable baseline:

A site was selected on the western bank of the Nile, nearly in front of the "Sphynx", and we took up residence in a house near the Great Pyramid. This house, very greatly enlarged, has now become the celebrated Mena Hotel [wrote David in 1912]; but in 1876 it consisted of only eight or ten rooms, and had never before been occupied. The Khedive supplied us with excellent servants, and installed us most comfortably. The training of officers and men had then to be undertaken. . . .

This was when everything began to go wrong. Try as he might, David could not seem to instil the proper methods in his Egyptian subordinates, and despite loud and lusty Scottish imprecations there was a continual series of misread drumheads, unreported knocking over of the rods, and general go-slow tactics. What David did not realize at the time was that he was being deliberately sabotaged. A clique of army officers felt that the job should have been theirs, and were doing their best to put the foreigner in a poor light.

The resourceful Scotsman, though, was not to be denied. Hearing that an American astronomer, Professor Watson from the University of Michigan, was in Cairo, he solicited Watson's help, and, dismissing all other helpers, the two of them completed the base by themselves. After that he resigned. Not much to his surprise he later learnt that his report containing all the measurements had been "lost" in Cairo, and "I have also since learned that the Arabs have uncovered the concrete piers and chiselled out the gun-metal blocks on which the lines defining the terminals of the various sections of the base were engraved."

After Dun Echt and Egypt, David and Isobel settled down for a short while in London. David was eager to put his new method of finding the sun's distance to a critical test, and he soon realized that an ideal opportunity would be forthcoming in 1877 when the planet Mars would make one of its closest approaches to the earth. Britain, however, would not be the ideal place from which to make the observations. In terms of miles per hour, a point on the earth's equator spins much faster than a point at higher latitudes, so in the course of a night, an observer at the equator travels much farther through space than one in Britain. Having the observer carried over a large distance by the earth's rotation was what David's method needed, so clearly he should choose some site at a low latitude.

In the autumn of 1876 he applied to the Royal Society for a grant of £500 to enable him to carry out an expedition to either of the islands of St Helena or Ascension for this purpose, promising to pay any additional expenses out of his own pocket. He was refused, though at the intervention of the Astronomer Royal the Royal Astronomical Society awarded him £250, and eventually (after the expedition had proved a success) the Royal Society did provide the additional £250.

One item would be essential: the heliometer. There were only two in the whole of Britain, and the Radcliffe Observatory at Oxford University would not part with theirs. But the amiable Lord Lindsay was happy to loan his to his erstwhile employee.

The instrument was set up in the rooms of the Royal Astronomical Society in Burlington House, London, both for checking and in order that David could demonstrate it in a lecture to the Society. And here the whole scheme very nearly came to a calamitous end before the expedition had even started. The working parts of the heliometer were small and very delicate, but they were mounted on a massive foundation for reasons of stability. In adjusting part of this David somehow managed to tip the heliometer, and the entire instrument went smashing into the floor. A bystander later recalled the scene:

And there—upon the front seat of the meeting room—sat Gill, his face buried in his hands, down which blood was trickling, as he had made an ineffectual clutch at the falling mass. He said something about everything being ruined—himself—the instrument—the expedition. It was painful to see a strong man so completely broken down. But it only lasted a minute or so: he suddenly got up and said, "Let us see what can be done." He instantly began his examination of the wreck, and asked me to go to Lord Lindsay and tell him of the accident. When I got back Gill had determined the extent of the damage, and decided upon the course to be taken. The vital portion of the instrument, the divided object-glass, had fortunately escaped injury, having been protected by the metal cap, which struck the meeting-room table, leaving a deep dent which is to be seen to this day [1916]. The eye-pieces with their tubes were ruined, but Gill would see Simms about them at once, and get them renewed. As we know, everything was done in time, and the expedition was an entire success.

Not that the heliometer was the only equipment. There were the usual barometers and thermometers, a transit instrument for determining latitude and longitude, and, of course, the inevitable chronometers, although merely five in number this time. Also an entire portable observatory for housing everything. Gill had finally settled on Ascension rather than St Helena after a scrutiny of the relevant weather records, and since there was no wood available on Ascension he had no option but to take an observatory with him. All told, with personal baggage, there were twenty tons to be transported.

David looked forward to it all with his usual exuberance. A strong contribution to his mood was the fact that this time Isobel would accompany him. Not that she made any claims to being an astronomer (David was once asked whether his wife understood all his esoteric astronomy, to which he cheerfully replied "Not a word, thank Gód!"), but they were too happy a couple to accept parting again for six months. Isobel, in fact, was to write a delightful book on their Ascension experiences, and as things turned out the expedition might well have been a failure had she not been there.

At least the voyages would not provide much trouble, for they would travel on the big steamships that plied regularly between England and South Africa. The south-bound ships called at St Helena en route, and the north-bound ones at both St Helena and Ascension, so it would be necessary to go down first to St Helena and then change ships to go north again to Ascension..

They sailed on the *Balmoral Castle* from Dartmouth on June 14, 1877, the critical date for the observations being in early September of that year.

It was a leisurely trip down, with a stop at Madeira, where they encountered that inevitable hazard of ship travellers:

We were boarded by a motley crew of gipsy-looking men eager to sell us photographs, shawls, feather-flowers and fruit. "Sell you a basket of cherries for six shillin' sare!" "Beautiful photograph—only three shillin'!" and some reckless, heavy-pursed colonist, whose box of presents from England is not quite full, lets himself be robbed in the most lordly fashion by these barnacles, while more patient passengers get plenty of cherries for one shilling when the last bell is ringing.

A few days later they sailed among the Canary Islands, coming within sight of Tenerife and its great 12,000-foot volcanic peak, which their old friend Piazzi Smyth had years before scaled to make astronomy's first observations in the infra-red part of the spectrum. Isobel was entranced.

It was a splendid transformation scene, and watching it I had forgotten my dinner, till the homely old Scotch stewardess interrupted my reverie with a plate of currant tart. I made some remark to her about the beauty of the scene. "Ay, ay," she said, "it's a big hill, but there's nae scenery in earth or ocean like oor ain Scotland."

July 1 saw their arrival at James Town, St Helena. It is a charming, sleepy little corner of the world in which little has happened since the death of Napoleon there; Isobel's description of the island during the week or so she and David were on it could apply almost unchanged to the St Helena I knew nearly a century later.

At the first peep of dawn I hurried on deck and saw, so close to the ship as to make me start, dark sterile rocks rising almost perpendicularly from the sea, and partly enclosing the bright blue bay in which we were anchored. At the bottom of a strange cleft in these fierce, fortress-looking crags, a quiet little town nestled close to the sea, filling up the lap of a valley scarce 200 yards wide.
   Here was the landing-stage, and just beyond, a row of dark Peepul trees fringed the shore, shading and cooling the cluster of low, white houses that we were so blithe to see. Besides these, little or no vegetation appeared. The great towering rocks were cold and bare. A long ladder of 600 steps sprang from the town up the steep western side, called Ladder Hill, and at the top I could descry some forts and the grim mouths of cannon.

Thanks to the good offices of Lord Lindsay, the Governor's secretary was at the landing-stage to meet the Gills, and soon they were installed for their week's stay at Government House. Gill was anxious to visit the site of an astronomical observatory established by an English astronomer Manuel Johnson some fifty years earlier. It stood at the top of Ladder Hill, and since Johnson had determined its longitude very accurately Gill could use it for checking his precious chronometers. Laboriously arrived at the top of the hill Isobel was indignant at what she found:

I say Observatory—alas! it is so no longer. Fallen from its high estate, it is now the artillery mess-room, and in the recesses formed for the shutters of the openings through which Johnson's transit used to peep, they stow wine-glasses and decanters, and under the dome they play billiards!

And since David did indeed set up instruments there for a few nights, one presumes there was some interruption in the billiard play, possibly after he had had a game.

On subsequent days they explored the island by horseback, although the precipitous landscape gave Isobel some bad moments (". . . the path had got so *ugly* that I shut my eyes and ignominiously grasped the pommel of my saddle. To make matters worse, I had been told that the pony I rode was a 'buck-jumper'. . . .") They passed by the weird 1,400-foot fingers of rock known as Lot and Lot's Wife, and explored Halley's Mount. Here two hundred years earlier, the second Astronomer Royal, Edmond Halley, in his younger days had maintained one of the earliest southern hemisphere observatories. David was delighted at being able to trace its remains. At the foot of the Mount is the tomb in which Napoleon was first buried, and not far off "Longwood," the house in which he lived and died. Sadly, the Gills found it almost derelict, although it has since been restored to a fine monument, kept just as Napoleon lived in it.

July 10 brought the *Edinburgh Castle* steaming into James Town Bay, and it was time for David and Isobel to say good-bye to "those grey beetling crags which hide so much softness and beauty." It would take only three days to transport them to a very different world.

What a sight [Ascension] was! The sun had been up some hours when we anchored in Clarence Bay on the 13th of July, and the "Abomination of Desolation" seemed to be before our eyes as we looked eagerly at the land.

A few scattered buildings lay among reddish-brown cinders near the shore—a sugar-loaf hill of the same colour rose up behind and bounded the view. We looked about in a sort of hopeless way for "Green Mountain", but it was nowhere to be seen, and we set it down as a fable—a mere myth. "Nothing green," we said, "exists, or could exist here." Stones, stones, everywhere stones, that have been tried in the fire and are now heaped in dire confusion, or beaten into dust which we see dancing in pillars before the wind. Dust, sunshine, and cinders, and low yellow houses frizzling in it all!

Is *that* Ascension?

Well, not quite; its coast presented a livelier scene, though one we would gladly have dispensed with. A black perpendicular wall of rock jutted out into the bay. It is on this rock that the "Tarter Stairs" are cut, and here we must land. But how? For this morning beautiful waves are dashing and crashing and splashing against the landing-place, or rushing past it in sportive fury. . . .

"The rollers are in!" . . . My heart grew heavy. But seeing H.M.SS. *Cygnet* and *Industry* in the harbour, I took courage, knowing that we should at least find refuge on board one of these vessels, and that we should not have to be carried on to Madeira,—a misfortune which has more than once happened to passengers roller-stayed at Ascension. . . .

We saw a gig put off from the *Cygnet,* and pull towards us. "An offer of hospitality," we thought, as we recognised the blue-jacketed oarsmen and their commander, whose acquaintance we had made at St. Helena.

"Can we land?" was our greeting to Capt. Hammick, as he came aboard. "Well, the flags denoting 'Doublerollers and Dangerous' are up at the pierhead, but the sea is going down, and I have permission for you to try it, if you don't mind wet feet." We didn't; so it was decided that I and the heavy baggage should be sent on shore at once, while the chronometers and more precious goods ! should wait for quieter times on board the *Industry.* . . .

I don't know how the heavy baggage liked it, but I certainly wished myself a chronometer more than once, when I saw, rising up behind us, a long wall of threatening water, and before us, the steep, dark rock, wet with spray. This feeling increased when we were within a few yards of the shore, and I found that we must get out of the strong trustworthy-looking gig, manned by its stout crew of English sailors, and trust ourselves to a little rickety cockle-shell, which at that moment was being baled out by two ebony-coloured boatmen. I thought, just then, they looked fiendish, and that I could see the baleful eye of a shark, certain of his prey, gleaming triumphantly through the green waves. . . .

"You may trust yourself with every confidence to these men," Capt. Hammick said to me; "they understand the rollers better than anybody else; they will not take you into danger, only you must be careful not to attempt landing until they give you the word."

For some minutes we kept dodging about, and once or twice were close under the steps; but we got no sign to stir, and were again and again driven back.

At last, there came suddenly a perfectly calm moment, immediately after an unusually heavy roller had tossed our little boat over its head, and we were again sculled under the rock in a twinkling of an eye. A rope was let down from above; David at once laid hold of it, and at the word "Now!" he jumped from the boat. I instantly followed his example, and thus gained a slippery footing on Ascension, with a somewhat palpitating heart and eyes smarting with salt spray.

At the top of the stairs they found Capt. Phillimore, the officer commanding the garrison on Ascension, and a group of his assistants waiting to welcome them. While David went off again to brave the rollers in seeing to his chronometers, Isobel was escorted "in a curious two-wheeled vehicle—which my conscience would not allow me to call a carriage, and which I was afraid to call a cart" to see the sights.

It was now nearly noon, and the dazzling sun shone with a pitiless glare on everything. I looked about me for some beauty to remark upon. But no! We passed great open sheds, piled roof-high with coals, square unsightly store-houses of various kinds, a creaking windmill painted red like a guillotine, and all thickly coated with a fine yellowish dust, into which our poor horse was sinking, hoof-deep, at every step as he pulled us up the gently rising ground leading from the wharf.

Having surmounted this we came upon a dreary flat, and still dust and ashes everywhere. Here we found facing us a neat little church; to the right the hospitals and marine barracks, with their two stories, interrupted a row of low-roofed, verandahed cottages, one of which I gladly learnt was to be our home. Beyond, were a few scattered, undecided-looking houses, with no character to speak of. We drove through these before beginning the ascent of Cross Hill, now rising straight before us. Again dust, ashes, cinders; a hill without beauty. . . .

Such would be home for the next half-year. Ascension, in fact, was just part of the bric-a-brac so casually accumulated by the British Empire in its heyday. There is a story of Gladstone, forming a cabinet for one of his innumerable Victorian governments, finding that he could get no one to take on the post of Colonial Secretary. So with a shrug of the shoulders he said he would do it himself, and "went upstairs to see where everything was on the map." Subsequently, at a dinner party he was asked by his neighbour where the Virgin Islands were. "Not the foggiest idea," he boomed cheerfully in reply, "although I suppose they are far removed from the Isle of Man."

Ascension had been known to mariners for centuries without any country wanting to acquire so grim and desolate a little island. Eventually the British established a naval garrison there at the close of the Napoleonic Wars for the sole purpose of denying it as a base to any possible French mission trying to rescue Napoleon from St Helena. By the 1870s, when the Gills were there, it served as a coaling station for the Royal Navy. In fact, the whole place was run strictly by shipboard rules, and its tiny population appeared on the books of the Admiralty as "the crew of the *Flora* Tender," thus permitting the administration of that life-blood of the British Navy, the daily tot of rum. Naval parlance was *de rigueur,* and Isobel had soon to accustom herself to speaking of getting dinner "under-weigh" in the "galley" and so forth. Captain Phillimore, however, ex-empted them from the standing ship's order of lights out at 10 p.m., "no doubt," says Isobel, "on the ground that an astronomer, being a species of lunatic, is not amenable to laws."

The main anomaly to this running of the island as a ship was the pres-ence of women, the wives and daughters of some of the senior officers. At times their presence tried Captain Phillimore's qualities of leadership to the full. Isobel relates how two officers' wives became embroiled in a bitter feud over which had the more senior rank and was thus entitled to a pew in front of the other at church. The problem was ultimately sub-mitted to the Captain for resolution, and he, in a decision worthy of Solomon, eventually pronounced: " 'Let age take the higher place,' said the wily Captain, 'and let the younger lady give way to the elder!' From this there was no appeal, and next Sunday, lo! both ladies were seated in the pew next to the door."

For a start Isobel and David had to sort out their twenty tons of baggage and get settled into their little cottage. Isobel, a true Victorian gentlewoman, was horrified at hearing how difficult it would be to find servants to do the cooking and wait at table, although two distinctly non-descript characters were finally hired for the job. Worse yet was the ques-tion of provisions. Fresh water, for instance, was so scarce that cooking had to be done in sea-water whenever possible, a quart of milk a week was considered a good ration, and fresh meat and vegetables very much a sometime thing. The naval canteen was usually out of anything one wanted.

No butcher! no dairy! no greengrocer! no fishmonger! only this wretched canteen, more full of flies than anything else. I got quite tired and hot with the frequent "No, madam, we don't keep it," or "Very sorry, but we are just sold out." My demands were modest, but they had to become yet more humble. . . .

I then turned to the open door with "Island Bakery" written over it—where a pallid baker stood at the threshold wiping the perspiration from his forehead. Evidently *he* made his bread by the sweat of his brow! "Can I have some bread?" I asked boldly, thinking there could be no difficulty here. "All served out for the night, ma'am." "Oh dear! and when do you bake more?" "The day after tomorrow!"

Isobel should not have worried. The bread, when available, proved most suited as fishbait, provided one had the strength to break it. Indeed, fish caught by the servants would become very much a staple diet.

Meanwhile David had set to work erecting the observatory. It was the kind of task he thoroughly enjoyed, going back to the days at Dun Echt when he would roll up his sleeves and get to work at bricklaying. "I wadna say what he may ken aboot astronomy," one of Lindsay's men had remarked, "but this I wull say, that he'd mak' a gran' mason."

Captain Phillimore had most courteously offered the use of the garrison's croquet lawn as a base for the observatory, the term "lawn" proving to be a euphemism for a blistering expanse of concrete slab, alongside which "a few withered aloes, with tattered dust-stained leaves, struggled for bare life." Still, as underpinning for the observatory it proved most suitable.

Despite the dreariness of the island itself, David and Isobel felt sure they would be rewarded when the opposition of Mars (the planet David was to use in his method of finding the sun's distance) arrived on September 5. To Isobel the skies over Ascension seemed beautiful beyond belief.

Sitting that first evening after sunset in the verandah which looked upon our novel croquet lawn, we could speak of nothing, think of nothing, but the beauty of the heavens. Though Ascension was barren, desolate, formless, flowerless, yet with such a sky she could never be unlovely. The stars shone forth boldly, each like a living fire. Mars was yet behind Cross Hill, but Jupiter literally blazed in the intense blue sky now guiltless of cloud from horizon to zenith; and thrown across in graceful splendour, the Milky Way seemed like a great streaming veil woven of golden threads and sparkling with gems.

Her elation was short-lived. It took a week or so to erect the observatory and get all the instruments in working order, and no sooner had this been accomplished than the unbelievable seemed to happen. Almost without exception every day was brilliantly sunny, yet no sooner had the sun gone down than clouds blanketed the sky, and night after night after night would be worthless for observations. This unbroken pattern seemed beyond all statistical likelihood, as though some malignant fate was determined to ruin the expedition.

> Oh! those weary weeks. Fearful of losing a single hour of star-light during the night, we watched alternately for moments of break in the cloud, sometimes with partial success, but more frequently with no result but utter disappointment, and the mental and physical strain, increasing every night, grew almost beyond our strength. What was to be done? There was the Observatory complete, the instruments faultless, and the astronomer idle. . . .

The torrid heat and dust of the daylight hours only added to their mood of sullen despair, and tempers began to flare at frequent intervals. It was Isobel who found the solution. Glaring around him one night, David made the chance remark that it was almost as though fate had determined to send the clouds over just the spot where the observatory stood; one could almost perceive clear sky off on each side. Isobel, far too irritable to sleep, immediately announced that she was going to take a long walk across the island and check that very point. David remonstrated. Go walking across the wild lava beds in the middle of a moonless night? But Isobel was adamant, so David went and woke up the more reliable of their two servants and ordered him to accompany her. What this worthy thought of being woken up and ordered to pack a picnic basket and go for a stroll with Mrs Gill at 3 am is unrecorded; presumably that if all Englishmen were mad, Scotsmen must only be more so.

> For the first mile I found the road very tiring—soft and unyielding, and bestrewn with loose lumps of clinker; moreover I had made the mistake of putting on low shoes. I chose them because they were thicker than any boots I had, not considering that the sand, or crushed cinder rather, would get inside and chafe my feet. Our next misfortune was the sudden rising of the wind, followed by the total eclipse of our lanthorn. But, happily, we had our bull's-eye to fall back upon, and by this light we proceeded. . . .

It was still very dark—the wind rose higher—the moon gave no notice of her coming, and the weird ghostliness of the little bit of surrounding that fell under the light of the bull's-eye, I shall never forget.

It was worth it. Even though the road soon gave out and they had to clamber over the clinker, a mile or two brought them to a spot where the skies were indeed clear. And now Isobel could see what was happening. By chance the observatory was directly downwind from the single great peak of the island. This, known as Green Mountain because the upper reaches of its 3,000-foot bulk were often shrouded in mist and produced the only greenery anywhere on the island, produced a long streamer of cloud in the night-time conditions of the steady trade wind that blew at that time of year. Elsewhere on the island the skies were clear.

There was nothing for it; they would have to give up their comfortable cottage and croquet lawn and move to an uninhabited part of the island if the expedition was to be a success.

David was despondent. It was already the end of July, the critical observing period only weeks away. How on earth could they move their twenty tons of equipment miles away across this ghastly roadless landscape, how would they live, what if a single stumble sent the heliometer smashing into the lava, could everything be set up and tested and ready again in just a few weeks? They would have to try.

Captain Phillimore was solicitous, and he and David spent some days hiking around the island searching for a likely spot. At last they settled on a coastal site, a little bay with a small white beach, its sides "bristling with clinker, erect and sharp as the quills of a porcupine's back." It would become known as Mars Bay. The delicate instruments and other small items would have to be carried very carefully across the lava beds, the heavier equipment they would have to try and land from the sea, something never before done there. It would depend on the terrible rollers.

Another writer of the period described the rollers thus:

One of the most interesting phenomena that the island affords is that of the rollers, in other words a heavy swell, producing a high surf on the leeward shore of the island, occurring without any apparent cause. All is tranquil in the distance, the sea breeze scarcely ruffles the surface of the water, when a high swelling wave is suddenly observed rolling towards the island. At first it appears to move slowly forward, till at length it breaks on the outer reefs. The swell then increases, wave urges on wave until it reaches the beach, where it bursts with tremendous fury. The rollers now set in and augment in violence until they attain a terrific and awful grandeur, affording a magnificent sight to the spectator, and one which I have witnessed with mingled emotions of terror and delight—a towering sea rolls forward on the island like a vast ridge of waters, threatening, as it were, to envelope it, pile on pile succeeds with resistless force, until, meeting with the rushing offset from the shore beneath, they rise like a wall and are dashed with impetuous fury on the long line of the coast, producing a stunning noise. The beach is now mantled over with foam, the mighty waters sweep over the plain, and the very houses at the town are shaken by the fury of the waves. The strong and well-built jetty of the town has once been washed away by the rollers, which sometimes make a complete breach over it, although it is twenty feet above high-watermark. On these occasions the crane at its extremity is washed around in various directions, as the weather-cock is turned by the wind.

Amid the tranquillity which prevails around, it is a matter of speculation to account for this commotion of the waters, as great as if the most awful tempest or the wildest hurricane had swept the bosom of the deep.

It is this sudden and quite unpredictable appearance of the rollers on the sunniest and calmest of days that is one of their most terrifying aspects. Isobel herself held to the romantic view that the rollers have their origins in the breaking off of stupendous icebergs from the cliffs of Antarctica, the splash setting off waves that travel out many thousands of miles ("I like this theory, and am glad to say that I have not yet heard it explained away.") In truth the rollers are a resonance phenomenon, produced by the action of geostrophic winds blowing across great reaches of open ocean.

But David was not much in the mood for debating the origins of the rollers; his question was whether or not they dare bring their heavy equipment around to Mars Bay by sea without getting caught by the rollers. He would have to take the chance.

On August 1 they dismantled the observatory and "plundered" the cottage of all belongings suitable for camping out on the clinker.

Next morning, as the sun rose, a rare procession passed down the coast. A steam-launch, with Captain Phillimore and David on board, towed along two well-laden lighters and a pinnace, and carried, moreover, quite a tail of little surface-boats, or "dingeys". The busy trade-wind had sunk almost into a dead calm, and the sea seemed still asleep, everything was in favour of an easy landing, and I felt hopeful, though anxiety made the hours seem long while I waited for news. I could neither read nor write, nor did idle musing soothe me, so I made believe to mend a pair of gloves, and ever after, when I wore them, I was wont to trace the anxious thoughts sewn in with every stitch. I threw them down in disgust, and, bidding patience good-bye, put on my hat and walked out into the noon-day sun.

"Do you see anybody coming?" "No!" That movement far off among the clinker is only the rising of the heated air, trembling over the burning stones. But at last, and sooner than I had any right to expect, there was the sound of wheels, and good news was brought to me. Everything had been landed without a scratch, the foundation of the Heliometer House was already laid, and the new harbour thus established, had been christened by Captain Phillimore "Mars Bay".

On the following morning another procession wended its way from Garrison [the town] to Mars Bay—this time by land. It consisted of sixteen Kroomen, bearing the Heliometer-tube, Transit and other instruments. The Heliometer box was lashed to a mast and set out on its perilous journey, borne on the shoulders of eight Kroomen—four in front and four behind. The other eight carried the lighter boxes and acted as a reserve. Strong stalwart fellows they were, looking like so many pillars sculptured in black marble.

Soon my husband followed in the cart, but what was his horror, on overtaking the procession, to find these faithless bearers had unswung the box, and were cooly carrying it on their heads. This mode of transport looked most unsafe, and he remonstrated, but to no purpose. "Krooboy must carry thing on him head—he no carry with pole—get tired." And so the trembling astronomer was fain to be content for the first part of the way, but when the plain was past and the clinker appeared, his patience gave way; he could bear it no longer. The box was accordingly lashed to the mast again, amid some grumbling at first, but it soon passed off, and a few kind words made the shining black faces as genial as ever. Then, with slow and careful steps, and with much laughing and chattering, the precious thing was borne over the rocks in safety, and when at last Mars Bay was reached, its tired guardian sighed out in his relief, "All's well that ends well."

But setting up house at Mars Bay hardly came under that heading. The Gills and their two servants had to live in tents pitched on the lava beds ("I am at a loss how to convey to anyone who has not seen it, an idea of what sort of flooring clinker makes.") Their water supply had to be brought by boat when the rollers permitted, and Sam, the lesser of the two servants, had to scramble across the clinker every day to fetch the food supplies. Food and water soon had a strong admixture of the ubiquitous yellow dust. Clouds of flies descended on them, and then, for the first time since they had arrived at Ascension, the skies everywhere grew dark and there was a torrential downpour of rain for several days. Their tents had been carelessly pitched with the doorways facing into the wind, and now the rain blew in and soaked everything and everybody. Isobel's relief at the departure of the rain was shortlived when as its aftermath hordes of "musquitoes" appeared. To cap it all, a week or so later David fell and injured his knee on the clinker, and then came down with a mild case of sunstroke. Isobel was beside herself.

But they had not come all this way to be beaten by such circumstances, and their Scots tenacity in the end won through. By the time the critical date arrived they were ready—shaky, perhaps, but ready.

Meantime the 5th of September has come. I could write no diary, and have not the slightest recollection of how I spent the day—unprofitably, I fear, in watching and waiting; finally bringing on a violent headache towards evening, which was less painful, however, than the excessive nervous excitement I was endeavouring to repress. Tonight Mars will be nearer us—his ruddy glare brighter than ever again for a hundred years, and what if we should not see him?

The sun had shone all day in a cloudless sky, but before sunset some ugly clouds rolled up from windward, and made me feel quite feverish. I could not rest, but kept wandering from tent to tent like an unquiet spirit; inwardly resenting David's exceeding calm, as a tacit reproof to my perturbation. There he sat, quietly tying up photographs, softly whistling to himself, as if nothing were going to happen, and then he actually smoked a very long pipe, with even longer and slower whiffs than usual. Of course it was affectation!

Six o'clock, and still the heavens looked undecided; half-past six, and a heavy cloud is forming in the south. Slowly the cloud rises ... and at last we see Mars shining steadily in the pure blue horizon beneath.

How slowly the minutes passed! How very long each interruption appeared! The wind was blowing lazily, and light clouds glided at intervals across the sky, obscuring, for a few minutes, *the* Planet as they crossed his path. But at last I heard the welcome note "All right, evening success", and then I went to bed.

This meant that the first group of observations, needed soon after sunset, had been a success. But for the method to work at all there must be another set made in the early hours before dawn. Would the broken clouds at least get no worse, or would they perhaps mass into a solid overcast and so ruin the night? Isobel couldn't sleep for worrying.

> I dared not go inside the Observatory, lest my uncontrollable fidgets might worry the observer, but sat without on a heap of clinker, and kept an eye on the enemy. Then David called out, "Half set finished —splendid definition—go to bed!" Just in time, I thought, for one arm of the black cloud was already grasping Mars.

Something at least would come of the night, but for it to be truly a success David would need further observations for the next few hours. Isobel struggled to sleep, but the excitement was too much for her and in the end she got up again. To her ecstatic amazement the sky was brilliantly clear from horizon to horizon.

> Mars now outrivalled Jupiter in ruddy splendour; Orion had flung abroad his jewels like hoar-frost; the Pleiades glittered in bewildering multitude. "Like fire-flies tangled in a silver braid", they shone with a soft beauty; and everywhere, above and around, myriads of stars dazzled the night.
> While my eyes drank in this beautiful scene, my ears were filled with sweet sounds issuing from the Observatory, "A, seventy and one, point two seven one; B, seventy-seven, one, point three six eight", &c. Let not any one smile that I call these sweet sounds. Sweet they were indeed to me, for they told of success after bitter disappointment; of cherished hopes realised; of care and anxiety passing away. They told too of honest work honestly done—of work that would live and tell its tale, when we and the instruments were no more. . . .

With the success of the expedition assured, David and Isobel were in dire need of rest and relaxation. As soon as they had got all the equipment back to their cottage they accepted Captain Phillimore's suggestion that they spend a week or two in another cottage on the upper reaches of Green Mountain. For months now they had seen no blade of grass, no flower, only the drooping dusty aloes to serve as flora. Isobel could hardly believe the transformation awaiting them on Green Mountain. Here were trees and grass and running streams, gorgeous flowers, and ("Oh, joy of joys!'') whole gardens of green vegetables. Happiness was complete when the cool and misty evenings brought them a roaring fire in the grate of their little cottage. Even the cockroaches couldn't spoil it. The final pleasure was the day they climbed to the very tip of the mountain and found carved on a rock there "Sic itur ad astra. Anno Domini 23 July, 1832"—"such is the road to the stars".

It would still be some months before the Gills would leave Ascension. David wanted to take the opportunity to do some additional astronomical work while there, and so it was that they spent Christmas of 1877 on the island. Like the British everywhere, they were not the least put out by the temperature being 89°F on Christmas Day, and looked forward to a solid dinner of roast lamb and plum-pudding. Unfortunately this failed to materialize, since their Krooman cook celebrated the day by getting blind drunk on rum. Being a decent fellow, and overcome with remorse, though, he made amends by serving them the dinner for breakfast next morning. David failed to see this as a peace offering, however, and not only was the cook dismissed but, after reporting to the Captain's cabin, found himself with thirty days' grog stopped and sentenced to tending bullocks on the clinker.

Wednesday, January 9, 1878 brought the mailship *Warwick Castle* steaming into Clarence Bay. It was time to leave. The laborious business of loading their baggage began, and

At sunset we embarked. Before steam was up we had time to arrange some comforts in our cabin, and to read the home letters which had come by the mail.

What a lovely evening it was, and how gloriously the crescent moon and Venus shone over the water and silvered the grim outlines of the land!

"For the last time," I said to myself, as the evening bugle sounded; and, before its echoes had died away, our "Six Months in Ascension" was a thing of the past.

Back home in Britain, with all the observations carefully examined and the final calculations completed, David found his name and reputation had been made. The expedition had been an enormous success, and his determination of the sun's distance proved to be one of the most accurate ever made in the nineteenth century; in fact, one that would remain the standard well into the twentieth century. Congratulations poured in from

astronomers all over the world, and the Royal Astronomical Society presented him with its Gold Medal, the Society's highest award for achievement.

David now found himself very much in demand as a lecturer and after-
dinner speaker, and one such occasion gave rise to a story against himself
that he enjoyed all his life.

Astronomers, for technical reasons, often speak not of the actual distance of the sun, but of an equivalent quantity called the solar parallax.
This is half the angle subtended by the earth at the sun, and is a very small
angle: about two thousandths of a degree. David liked to impress his lis-
teners with how small an angle this is by telling them that it was the angle
presented by a British threepenny-bit a hundred miles away. This angle
had to be measured with an accuracy of better than one percent. At one
of his after-dinner speeches he alluded to this coin a hundred miles off
several times, and when the master of ceremonies rose to thank David for
his speech, he gravely remarked that clearly astronomy was an esoteric
subject, but only when he had heard their speaker's accent did he realize
it would take a Scotsman to be so concerned about threepence a hundred miles away.

The only nagging question amid all the acclaim was What next? David
must very soon find some other appointment. A very promising avenue
seemed to open up when the directorship of Oxford's Radcliffe Observatory fell vacant. The Radcliffe had one of the largest and finest of heli-
ometers, and David now had a reputation as one of the most skilled of all
observers with that instrument. But, possibly for political reasons, he did
not get the job. Instead it went to Edward Stone, at the time director of
the Royal Observatory at the Cape in South Africa, where for the last
nine years he had been successor to Thomas Maclear.

But this in turn left the directorship at the Cape open. Gill was very
hesitant about even applying for the post after his rebuff by Oxford, but
unknown to him his old friend Lord Lindsay was at work behind the
scenes pulling the necessary strings. The initial result was a rather mystifying exchange:

On the 10th of February 1879, as I was leaving my club, I met the
Hydrographer of the Navy, who said, "Let me congratulate you."
I replied, "On what?"—when he said, somewhat hurriedly, "Oh,
never mind; perhaps I am wrong." I drove home hastily to make
inquiry, and found a letter from Lord Crawford, addressed to me as

Astronomer Royal at the Cape, congratulating me on election to that office. It was evident that Lord Crawford had written in support of my candidature, and had received an early intimation of the Admiralty decision.

So it was that the Gills would travel again, and, in the long tradition of Henderson and Maclear, see what they could make of the Cape Observatory.

There were all the usual dire warnings. One acquaintance wrote to say that "South Africa has become the grave of great reputations," and accordingly wished them all the luck he clearly felt they would need. Piazzi Smyth, on the strength of his early years at the Cape as assistant to Maclear, fired off practical advice worthy of Henderson:

> In the way of entomology, I never saw a real disgusting B-flat, as a musician said, except on a parcel brought to the Obs^y out of Cape Town: but the lively little F-sharp is to be kept in order by nothing but abundant washings down with soap and water; and therefore, no carpets! [These are presumably whimsical references to roaches and fleas.] But there is another flat thing they call a Bushfly, a creeping flat brown affair, who in the summer contrives to get upon you in your walks, and if you do not look sharp he begins burying himself head-first into some convenient place for him between your shoulders and very inconvenient for you to get at him. Husband and wife may then be of inestimable service, for if you get hold of the body of the creature you must pull gently only, or the head will come off; and being left in your skin will make the cure rather worse than the disease.
>
> Of reptiles, you must be forewarned of the snakes. Occasionally a poisonous cobra is met with; and occasionally also a puff-adder which is worse, for it will pursue to bite, as well as bite when pursued.

And the final advice came from the Astronomer Royal, Sir George Airy: "Promise me, Gill, not to become a dress-coat astronomer," which probably meant that Gill was not to fall prey to the social life of Cape Town, but to get on with the serious business of astronomy. Gill would succeed admirably at both.

David had hoped that there would be some overlap at the Cape with his predecessor, Edward Stone, for after all the advice he had received he did not much care to arrive in this strange country and immediately find himself alone in his directorate. But—

My wife and I reached the Cape on the 26th of May, 1879, after a voyage of twenty-four days in the R.M.S. *Taymouth Castle.* Mr Stone welcomed us on arrival and, to my great regret, informed me that he must sail the following afternoon for England; thus there was little time to discuss with him the past and future policy of the observatory.

The past policy of the observatory seems to have left much to be desired, and the Gills were despondent at the many signs of decay and neglect they found around them. Living quarters, grounds, and instruments alike would need a good deal of refurbishing. The fault for this was not to be laid entirely at the door of Mr Stone; he, like his predecessors going back to the Rev. Fallows, had found the task of persuading the Admiralty six thousand miles away to spend the necessary funds on their observatory to be beyond him. What was needed was a really tough, persistent director; someone, in fact, like David Gill.

Within weeks of their arrival the Gills found themselves involved in a sad occasion, as David detailed in a letter to the Astronomer Royal.

> Royal Observatory,
> Cape of Good Hope.
> 1879, July 14.

My Dear Sir George,

I write to tell you that Sir Thomas Maclear died this morning. He has been confined to bed since my arrival in the Colony, but it is only in the last fortnight that his friends thought him to be dangerously ill. I have seen him three times. On the last two occasions he was very weak but full of pluck, and declared that he was quite well. The first time I saw him he was full of anecdote and fun, and his intellect was as clear and fresh as possible.

He impressed me as a man who must have been full of restless energy, a man of many sympathies, full of heartiness, and full of his work too. His Observing books bespeak the man. There is a scrupulous care about the notes, a constant personal attention to every detail, and an amount of personal labour in observing which few men have equalled. . . .

Sir Thomas is universally respected and loved in the Colony. We bury him Wednesday, beside his wife in the Observatory Grounds, near the spot where Fallows lies.

> Believe me,
> Sincerely yours,
> David Gill.

And now David must get down to the task of persuading the Admiralty that improvements had to be made to the observatory. As far as living-quarters and the grounds were concerned, he found the Admiral commanding the naval base at nearby Simonstown to be most co-operative, and before long teams of men were at work on these aspects. But when it came to the astronomical instruments he would have to deal directly with London, and soon he was deeply involved in the mystifying machinations of the Civil Service.

At one stage he was pleading for the grand sum of £2500 to buy a new telescope, to erect a building to house it in, and to extend his diminutive staff. This produced a stiff rebuke and warning: "My proposal was condemned as too costly, and I was warned 'of the costly character of all proposals emanating from the Cape, and to show more care in future.' " Scarcely had Gill digested this rebuff than it was followed by a letter from another official in London informing him that £3000 was granted for the proposal.

The minds of minor officials who failed to appreciate the local conditions were a never-ending source of irritation. When David wanted to improve the functioning of his transit telescope he applied for funds to erect two collimation markers. These were fairly simple concrete pillars, set on bedrock at some distance from the telescope, which would contain index marks to enable the astronomer to measure any displacement of his transit telescope from the exact meridian. London replied that according to their advice astronomers in Europe generally relied on observations of church steeples or edges of prominent buildings for this purpose, and why did Gill not do likewise and save the expense of special markers. Sitting out on the African veld, Gill's exasperation at receiving this drove him to write out a detailed mock-serious proposal for funds to build two cathedrals, until prudence got the better of his temper.

But David had at least one major advantage over his predecessors in dealing with officialdom. Whereas the earlier directors had found themselves practically stranded at the Cape, David could much more easily get back to London to deal with things in person. Maclear, contemplating the horrors of a two-month voyage in each direction by sailing-ship, had made only one return trip to England in the forty-odd years of his tenure. Gill would go every two or three years, a comfortable three-week voyage by steamship. And dealing with the commanding personality of Gill, his voice booming and fist thumping the desk, was a very different proposition from dealing with Maclear, writing letters of supplication from six thousand miles away. Gill once reported to his wife in some wonderment that at one of these confrontations he had actually been sworn at by some official. "What did you do?" asked Isobel. "Swore right back," said David.

Nowhere was Gill's ability at handling officials better demonstrated than in the affair of the heliometer. From the day he had set foot at the Cape it had been David's cherished ambition to count a good heliometer among his observatory's equipment. In due course the proposal was made, and as always there were endless delays and stalling on the part of the Admiralty. But a Gill denied his heliometer was something to be reckoned with, and eventually he showed up in person to find out what was going on. His assistant recounts the tale:

> On arrival at the Admiralty one morning he found that the question had passed from the Hydrographic Department and that before reaching the Treasury would pass through many hands and might be settled in about three weeks. After careful enquiries on general procedures he traced the documents and cheerily interviewed the official in whose hands they were, and explained the importance of the instrument and its uses. Thanked for his kindness in calling he was told the request would receive early attention and would probably be out of that room in a week or two; but Gill insisted that it was essential that it should be through all Departments of the Admiralty and sanctioned by the Treasury to enable him to announce to the Astronomical Society Meeting that evening that the Government had sanctioned the purchase of the instrument. After suggesting that the official could write his brief minute at once as well as a week later his views prevailed, the minute was written, and he was entrusted with the documents for conveyance to the officer who was to deal with them next. The process was repeated and he hied him to the Treasury where he added to his former plea for haste the example of the businesslike way the Admiralty had dealt with the matter. The Treasury people humoured him, but the last man urged the utter impossibility of final Treasury sanction as the Financial Secretary was not in his office. Enquiry as to his whereabouts proved him to be at the House of Commons; so Gill hastened there and after explanations the Secretary agreed to the provision of the heliometer; and a very happy Gill drove at once to the R.A.S. and made his announcement.

The arena of local politics in Cape Town also often felt the impact of David Gill. He was undoubtedly regarded as an important man in many affairs at the Cape, and not only moved in the highest social circles, but was a personal friend to several successive Governors who frequently sought his opinion on matters under their jurisdiction. In dealing with lower-level politicians, however, he did not get on nearly so well, as witness the beginnings of a letter from one such injured official:

The Treasury,<br>
Cape Town,<br>
July 8, 1899.

My Dear Gill,
    Thank you for your kindly note. You seem to know nearly as much about politics as I do about astronomy upon which, however, I seldom give my opinion. . . .

The staff at the observatory long treasured the story of Gill and the Railway officials. As Gill succeeded in rapidly expanding the staff, more and more of them lived away from the observatory itself. Then, as now, a railway commuter line linked Cape Town to its string of southern suburbs, and a special stop had been put in for the observatory staff, from which they would walk a mile or so to their work. At one stage the Railways decided that this stop was uneconomical, and informed Gill that henceforth no trains would stop at Observatory (as the station was, and is, called). That was most unfortunate, wrote Gill in reply, because it would mean that his staff would now take longer walking to work from the next station, and coincidentally that was just the time they needed for supplying the Railways with their accurate timesignal. Trains have stopped at Observatory ever since.

How strongly David could feel over political matters was shown in 1896 when he learnt that he was to have a knighthood bestowed upon him. He was, of course, pleased, until he discovered that the same order of knighthood was to be granted several Cape politicians whom he regarded as utterly worthless. He promptly wrote to say he would flatly refuse the knighthood rather than belong to the same order as such people. Instead it was diplomatically arranged for him to receive a different order of knighthood, and so he became eventually a K.C.B. It would now be Sir David and Lady Isobel.

Far from South Africa being the grave of his reputation, David's renown as an astronomer rapidly grew during his years at the Cape. He was already famous for his work on the sun's distance, but three years after his arrival in Cape Town there occurred an event that would make

*The Cape Observatory staff in 1879 (from* A History and Description of the Royal Observatory (Cape of Good Hope), *HMSO, London 1913)*

him even more famous. The year 1882 produced a singularly bright and beautiful comet, and David determined to try and photograph it. He couldn't use one of his ordinary telescopes because they took in only a very small section of the sky, and the comet was spread across many degrees. An ordinary portrait camera was needed, and indeed several people had already tried photographing the comet in this way, but the problem here was that the comet—like the sun and moon and stars—moved across the sky from east to west, so that pictures taken with a fixed camera were blurred unless the exposure was comparatively short. David had the idea of using a portrait camera attached to the side of a telescope; in this way the telescope would provide the necessary tracking motion while the camera took a long exposure. He would need someone experienced with portrait cameras, however, so "I accordingly called upon Mr Allis, a photographer in the neighbouring village of Mowbray, of whose skill as a photographer I had previous experience. No sooner were the objects of my visit explained to him, than he volunteered all necessary aid, and entered into the work with heart and soul."

The picture would be more successful than they ever dreamt. The comet appeared spectacular enough, but it was not that which struck David in looking at the picture. It was the enormous number of faint stars that appeared everywhere in the background. David suddenly realized that here was the ideal way to construct star maps: using wide-angle lenses and long exposures to record faint stars over great areas of the sky.

To produce catalogues of the stars shown on the maps, though, would require a great deal of laborious measurement, and here David was fortunate in acquiring the co-operation of the great Dutch astronomer Jacobus Kapteyn. At Gronigen in Holland, Kapteyn got little clear weather for taking astronomical pictures himself, but his laboratory was well-equipped for measuring pictures taken elsewhere. So if David would get the photographs, Kapteyn would arrange the measurements. Even so it was an immense undertaking, and before long others had been brought into the scheme, until eventually it grew into one of the most wide-flung international projects ever undertaken in astronomy. In revised form, and with far more sophisticated equipment, it continues to this day.

David did a great deal of other astronomy at the Cape, particularly with the heliometer, but it was this photographic scheme that set the seal on his fame. Again he received the Gold Medal of the Royal Astronomical Society, as well as many other medals and honours from foreign countries (they take twenty lines to list under his name as author of a book he wrote). He was soon an intimate not only of the great astronomers of Europe, but also of such men as Edward Pickering, Simon Newcomb, and the young George Ellery Hale, who were then busy pushing the United States into world-leadership of observational astronomy. Newcomb in particular became a close friend, and, through Gill, found himself in the role of newspaper reporter interviewing the Zulu Chief Cetawayo in South Africa in 1883. Almost the ultimate measure of David's esteem came when the British national astronomical hero, John Couch Adams, died, and Cambridge University offered David, who had started as a watchmaker and never earned a university degree, the Chair of Astronomy there. However, he preferred observing in sunny South Africa to cloudbound Cambridge, and declined the offer in favour of his friend Sir Robert Ball.

Life at the Cape was almost as full with the famous of other fields. The Observatory, like observatories everywhere, had always been a drawing card for visitors to the Colony, but at the Cape they often came on business too. Many of the great African explorers and military men of the nineteenth century had associations with the Observatory in the rating of their chronometers and checking of their survey equipment. Maclear, for instance, had made quite a friend of the missionary-explorer David Livingstone, and in Gill's day the trend continued.

One who came for different reasons was the strange mystic General "Chinese" Gordon, at one time commander of colonial forces in South Africa. Gordon was something of a forerunner of such people as Lawrence of Arabia or Orde Wingate in Burma, a man of unorthodox yet astonishing leadership and organizational abilities, who had become famous for his exploits in China, India and Egypt. But he was a curious mixture of religious zealot and secret alcoholic, strange in the folkmeaning of that word, and his visits to the Observatory were sometimes tense. Isobel would be asked for a bible and a copy of Milton's *Paradise Lost* so that Gordon could show the Gills what he believed to be a clear and logical deduction of the precise geographical location of the Garden of Eden. No one was convinced, but nevertheless both he and the Gills thought highly of one another.

Someone with whom David had more practical dealings was Cecil Rhodes. Rhodes had opened up what would become Rhodesia and other territories to the north, and Gill, who was heavily involved in geodetic surveys, worked hard to convince Rhodes that he should finance similar schemes in the new territories. Just as Rhodes wished Africa to be British from Cape to Cairo, so it was David's dream to see surveyed the longest north-south land arc in the world, running from the southern tip of Africa to the northern tip of Scandinavia. In his day it would have had great scientific use, although in the event it was not completed until the middle of the twentieth century, when its usefulness was much diminished.

In pursuance of this, David would often go to visit Rhodes at his magnificent home of Groote Schuur (now the official residence of South African prime ministers), located a couple of miles from the Observatory at the foot of Devil's Peak.

> One of the most delightful things about [Rhodes] was his joy and delight in the beauty of his surroundings. He would sit under his verandah at teatime looking up at the great mountain before him, and ask you passionately: "Is there anything more beautiful in the whole world?"
> He would turn upon you suddenly and say, "Did you ever realize what a privilege it is to be an Englishman?" And if I mildly suggested that it was better to be a Scotsman, he would say, "Ah, man, that is the same thing."

Convincing Rhodes of the need for the survey proved difficult:

> Mr. Rhodes said to me, "Yes, that is a fine scheme—a fine scheme; but you must remember that I must first of all provide something in the way of roads and bridges to facilitate communication, and when we have got so far in that direction I will support your survey." Then, turning to a map of Africa, he said, "Look here, a man requires two things to enable him to do a great work in the world; these are, first imagination, and next grit. The French have got imagination, but we have mostly the grit without the imagination. . . . To those who have got imagination and grit everything will come. Now, good-bye, I won't forget my promise."

But even when Rhodes had given the go-ahead David found Rhodes' minions in London procrastinating. He called again at Groote Schuur (a Dutch name meaning literally "Great Barn") to register a gentle complaint. Rhodes in exasperation turned to his secretary, saying, "Take a telegraph form and write: I have promised Sir David Gill that I will carry out his Arc of Meridian. Tell them to *find* the money! The rest is all red tape." Then, turning to David, "Fine thing, money." "Finer thing astronomy," replied Gill. "Too damned expensive," said Rhodes.

Through Cecil Rhodes David and Isobel became friendly with a young and up-coming writer and his American wife: Rudyard Kipling. The Kipling family, after an unpleasant lawsuit developed over their home in Vermont, spent a good deal of time at the Cape, and they were frequent guests at Observatory dinner-parties.

Also through Rhodes David established a lifelong friendship with Earl Grey. This developed through the years when Grey often passed through Cape Town as administrator of Rhodesia, but it continued long after Grey left to become Governor-General of Canada. It seems that Grey kept up a continuing competition with the Canadian prime minister, Sir Wilfrid Laurier, as to who could produce the best Scotch jokes, and since Gill was renowned for his store of such jokes, Grey was forever writing for more ammunition.

Government House,
Ottawa,
February 14, 1905.

My Dear Astronomer,

A thousand thanks for so kindly writing to me from Johannesburg to tell me about my boy. He writes me excellent letters which lead me to believe he is both interested and happy. . . .

Here is a story which will amuse you. A Custom House officer put the usual question to a Scots lady the other day on arrival at New York, as to whether she had any dutiable goods. "No, nothing but wearing apparel," she persisted, and showed some indignation when the Custom Officer, distrusting her word, proceeded to open her box and rummage to the very bottom. With triumph he pulled out from below her dresses two big magnums of whisky, and holding them by the neck asked the lady what she meant by saying that she had nothing in her box but wearing apparel. "I stated what was the truth," said the lady, "for you hold in your hands my husband's night-caps!"

Can you send me back a better one which I can tell Sir Wilfred Laurier, whose story this is?

I am much distressed that you are not able to give me a better account of your delightful wife. Please give her every assurance of my continued devotion.

When you have time please dash me off a line, for I enjoy keeping myself in touch, as far as possible, with South Africa.

I remain,
Yours ever,
Grey.

The turn of the century saw South Africa caught up in the flames of the Anglo-Boer War, a war that had an extraordinary effect on David Gill, even though he was never involved in any of the actual fighting. Knowing Rhodes, the colony's Prime Minister, and being a close friend of Sir Alfred Milner—the Governor, and a man of unusual administrative ability and wisdom—David could foresee how unpleasant a war this would be, and how dire its repercussions. Many of the British at the Cape felt this, unlike their jingoistic brethren in England, although Rhodes, of course, would largely be the war's shadowy architect. Britain had not fought a major war against a white nation for almost half a century, and the years of gunboat diplomacy and lightning bolt dispatches against unsophisticated tribes had lulled her into a completely false sense of overpowering strength. The sight of the world's most powerful nation setting out to smash the tiny populations of two small republics, and having the greatest difficulty doing so, was as unedifying as would be the Vietnam War seventy years later. It was a war which in retrospect marked the beginning of the British Empire's decline, a war whose echoes have not completely died out in South Africa to this day.

It began in earnest in 1899, and David would often be in despair during the next few years. On one occasion he spoke feelingly of a vivacious dinner party he and Isobel attended, when, a week later, literally half the men around the table had been killed.

In 1901 the British Government, presumably to bolster morale, sent out the Duke and Duchess of York, later to become King George V and Queen Mary, on a tour of the British colonies. The Admiral at Simonstown invited David to join him on his trip to Durban on the occasion of their Royal Highnesses' arrival. David, in fact, had already met the Duke in 1881, when he and his brother had arrived at the Cape as youthful midshipmen on a Royal Navy vessel. ("Lord Charles Scott, their Captain, brought them out to dinner one evening at the Observatory. They made great fun of making the Dome go round, and specially enjoyed a forbidden cigar when the Tutor was star-gazing. . . .")

At 10 o'clock the tug with the Duke and Duchess and suite landed. The Admiral presented me, when the Duke said I did not need an introduction as he had dined with me twenty years ago at the Observatory—"and a very jolly evening we had," he said, catching my eye.

The war ended in 1902, and David recalled the Observatory scene:

It was only 10 a.m. I told my young men to try and work until noon, and then go—but they couldn't, and I couldn't—and at 10.30 I said, Go and hoist every bit of bunting—and get out all the guns you have, and fire a royal salute and come in to me. And this done they all came into my room, and some 25 of us drank the King's health and Roberts' and Kitchener's and Buller's and French's—in my best champagne—and sang God Save the King—I tried to make a speech and could not—and we all went home or into town, to shake everyone we met by the hand.

One of the participants recorded something David could not:

He rose to speak. Not a word could he succeed in uttering. After we had waited through two minutes of expectant silence, he sat down at the table with his face between his hands, and sobbed. It was the most eloquent speech he ever made.

In 1908 David turned sixty-five, which was the mandatory age of retirement, but in fact he had already retired from the Cape Observatory by then. In 1906 his health had begun to fail him; there were dizzy spells and twice he unexpectedly fainted. On medical advice he applied to the Admiralty for an early retirement, and this the Admiralty reluctantly granted, effective from 1907. The truth, however, which David confided only to his closest friends, was that he was far more worried about Isobel's health than his own, and was really using the latter as an official excuse to retire to England. Isobel had been seriously ill on several occasions over the previous ten years, and David very much feared that another summer or two of the African sun would be too much for her. So worried did David become that in the end he took some of his long-accrued leave in order that they could depart even earlier. Officially, though, he said otherwise:

The home where a man has spent the best twenty-eight years of his life amidst stirring national events, in surroundings rich in every natural beauty, and with a life full of the work that he loves, cannot be quitted without a sharp pang of regret. But the decision had been fully considered. I realised that the most vigorous years of my life were behind me, and that the time had now come to turn over to younger hands the work in which I had long rejoiced.

Therefore, on the 3rd October 1906, my wife and I bade good-bye to our beautiful home, taking with us treasured memories of the many happy years spent under its roof, of the loyal and cordial support of my fellow-workers, and of the many other good and true friends we left behind.

The retirement years were quiet ones spent in a small house in Kensington, London. David found himself in somewhat straitened circumstances financially, and was forced to take on writing and lecturing tasks he would rather have avoided. More to his liking was the writing of a monumental history and description of the Cape Observatory, and his portly figure and booming voice were familiar at most meetings of the Royal Astronomical Society.

Despite his fears for Isobel's health, it was he who died first. Although vigorous to the end, he caught a chill while attending the funeral of his old friend Sir Robert Ball, and this very quickly turned to pneumonia. Within two weeks he was gone, aged seventy, on January 24, 1914. At least he was spared the horrors of the World War.

Isobel survived him by five years, dying on August 31, 1919.

They are buried side by side in the ruins of the fourteenth-century Cathedral of St Machar in their beloved Aberdeen. The site is very simple, as Isobel described it after David's funeral: "The grave is turfed over and no flowers are placed upon it—only the chaplet of laurel leaning against the old wall."

*Edward C. Pickering in the later years of his administration (by courtesy of Harvard College Observatory)*

*Solon Irving Bailey, acting director of the Harvard College Observatory, 1919-1921 (by courtesy of Harvard College Observatory)*

# "WE HAVE TWO OR THREE REVOLVERS. . ."

## *Solon Bailey in South America*

The barren hills swept down to the water, and the snow on their upper reaches added a bitter chill to the already piercing wind. The two men on deck huddled more closely into their overcoats and stared in amazement at the *cholos* on the quay at Puno calling farewells to their friends. The women went barefooted, barelegged, and even barebreasted, seemingly impervious to the cold, and the men, while they wore tight, brightly-coloured woollen caps on their heads, were barefooted too, with their trousers curiously slit up the backs of the legs.

Marshall, shivering, took stock. Monday, November 28, 1889. It was the coldest summer weather he had known, and yet he was in the tropics. And still, he thought grimly, with over a hundred miles to go across the lake. He breathed deeply, for here at almost 13,000 feet above sea-level the air was thin.

At last, with a cheerful blast of its horn, the *Yapurà* set off across Lake Titicaca. The little 60-ton vessel was an ex-gunboat which had been laboriously hauled up, piece by piece, on muleback across the high cordillera from the coast. Now its captain, Senor Salaverry, came out on deck to see how his American passengers were faring. Hearing of their interests, he kindly changed the course of his boat so that they would pass close to the islands of Titicaca and Coati. It was said, he explained, that on Titicaca the first of the Incas descended, sent by his father the Sun to instruct mankind. And as the boat passed up the east coast of Coati there became visible the Temple of the Virgins of the Sun, its glory long past, but still well-preserved.

Captain Salaverry went to settle some complaint of his engineer about their fuel reserves of dried llama dung, and the two brothers hunched down against the wind. Beyond the distant hills rose the magnificent Cordillera Nevada, and Solon remarked that they must be the finest mountain range in all the Americas. Marshall nodded mute agreement. He was beginning to feel he had seen enough mountain scenery to last him a lifetime. If only the damn wind would stop. They passed from Peru into Bolivia.

At seventy-four there's no doubt that Uriah Boyden had become something of a codger. On several occasions during his life he had given considerable sums of money to Harvard University for such purposes as the buying of physical science books for the library—not in itself a particularly eccentric act, yet when the Harvard College Observatory approached him for a contribution towards the observatory's expenses he behaved in a most peculiar manner. One of the committee that had made the solicitation explained:

> [Boyden] had promised Mr Wm Amory & me to send us a check for five hundred dollars but declined to put his name on the subscription paper. When nearly all the money was raised I asked for his check —he declined saying in a very coarse manner that we were getting money under false pretenses. . . .

Nevertheless, when Boyden died the following year, 1879, it turned out that the crusty old engineer had left virtually his entire fortune—some $230,000—to astronomy. Not, it is true, specifically to the Harvard Observatory, but in trust until some astronomical institution could convince the trustees that an astronomical observatory would be built on a mountain peak "at such an elevation as to be free, so far as practicable, from the impediments to accurate observations which occur in the observatories now existing, owing to atmospheric influences."

Mr Boyden's heirs not unnaturally saw this as eccentricity run rampant, and there ensued a legal battle of considerable proportions. But when the smoke had cleared the will was held valid, and the money really was available for the building of a mountain observatory.

The Harvard College Observatory, like many another, was perennially short of money, but in 1876 it had had appointed as its director a zestful thirty-one-year-old Bostonian, Edward C. Pickering, who was to prove something of a genius at the delicate art of soliciting funds from private donors. Now, in an early example of his talents, he turned his attention to the Boyden Fund. It took some time, waiting for the settlement of the litigation over Boyden's will, and then having to compete against other institutions eager for the money; but he succeeded. On February 14, 1887, the trustees of the Fund formally turned it over to the President and Fellows of Harvard College.

Pickering's talents ran to much more than mere fund-raising. He was one of the early American astronomers who could foresee the future importance of astrophysics as distinct from the older classical astronomy. He wanted to find out about the nature of the stars themselves, rather than merely their motions and positions. To do this really well he realized that the observations must be made at sites away from the all-too-often cloudy eastern United States. A mountain observatory at a clear weather location would be very desirable. In fact, Pickering, who thought big, really wanted two such observatories: one in the northern hemisphere to study northern stars, and one in the southern hemisphere for southern stars.

When the news got around that Harvard had won the Boyden Fund many of Pickering's friends wrote in with suggestions for possible sites. Samuel Langley enthused over the fine (astronomical) conditions that he had found atop Pikes Peak during a solar eclipse expedition there; David Gill announced that the interior of South Africa would offer excellent choices; while another suggestion was the 14,000-foot peak of Mauna Kea in Hawaii.

But travel was slow in the 1880s, and Pickering set his sights on locations as close to home as possible. Since the Pacific coast of South America lies almost due south of Boston, he began to think of the Andes for the southern hemisphere site, and the South-Western United States for the northern observatory.

It very soon seemed that Harvard's northern observatory would be in California near Los Angeles. James Lick, a millionaire and eccentric among eccentrics, had recently been dissuaded from his original intention of building a monument to himself in the form of a pyramid, larger than the pyramid of Cheops, in downtown San Francisco, and instead to provide a monument in the form of a mountain observatory. The Lick Observatory was put in the hands of the University of California, and built on Mount Hamilton a short way south of San Francisco. This location engendered considerable jealousy among a group of citizens in Los Angeles and Pasadena, who came to feel that if their northern rivals had an observatory so should they. They soon found a location, Wilson's Peak—later known as Mount Wilson—outside Pasadena, but they needed astronomers and financial backing to proceed.

So it was that Harvard, with money and astronomers, made a natural partner to the citizens' group in Southern California. Pickering sent out observers to check on the astronomical qualities of Mount Wilson, and was pleased to have their report that it was "an astronomical paradise." The way to establishing an observatory now seemed clear, and to make a formal tie to Southern California, the University of Southern California was brought into the picture. But negotiations were delicate; the Californians were somewhat reluctant to allow Bostonians in on their deal, while Harvard could foresee difficulties in the running of a shared observatory. The upshot was almost ludicrous. After Pickering had approved a draft version of the legal agreement, one of the Harvard lawyers decided to make a few minor revisions to it. The final version was sent off without Pickering seeing it, so not until later did he discover that in a moment of New England parochialism the lawyer had put down the agreement as being with the University of California instead of the University of Southern California.

The citizens of Los Angeles were outraged. Harvard was trying to sell them out to their northern rivals again! Immediately they withdrew their support, blocked Harvard's way, and the whole fragile agreement collapsed. It was a tiny mistake that would in the end cost Harvard dearly. For, in other hands, the Mount Wilson Observatory would one day become one of the most important in the world.

Harvard's southern observatory, despite a slow start, would prove much more satisfactory. From a study of such weather records as were available from the South American republics, Pickering decided that a starting point of any search for a suitable site should be the town of Chosica in Peru. Next he would have to send down a small expedition to begin the laborious examination of individual mountains until the best one could be found. For this task he chose Solon Irving Bailey.

In 1888 Bailey was thirty-three and already well known at the Harvard Observatory for his dogged determination in getting things done. Four years earlier, while the headmaster of a school in Tilton, New Hampshire, he had written to the President of Harvard asking to be admitted to the Observatory in order to study for an MA degree. The Harvard Observatory, however, was not in the educational business; at the time it was purely a research institute. Students who wished to study astronomy did so under the aegis of James Peirce, Perkins Professor of Mathematics and Astronomy, from whom, no doubt, they received thorough instruction in classical astronomy. Bailey wanted training in the new astronomy, but Pickering put him off, suggesting that he apply to another university.

But such was the young man's persistence that within another couple of years he had wheedled his way into the Harvard Observatory as an unpaid assistant. Here he would do forty hours' work a week purely for the privilege of learning techniques, while still having to cope with Professor Peirce's formal requirements for his MA degree. His capabilities proved the equal of his enthusiasm, and Pickering rapidly warmed to

Bailey. Before long Pickering was paying him a miniscule salary, and suggesting to Peirce that much of what Bailey was doing at the observatory might be counted in lieu of course work towards his degree. So it was that largely through his own efforts Bailey proved his worth and in 1888 was awarded an MA degree. Now, starting at the beginning of 1889, he would spend two years in South America looking for a suitable observatory site. As his assistant he chose his brother Marshall (a professional photographer, always rather quaintly referred to in Solon's reports as Mr. M. H. Bailey), and to round out the party there would be Solon's wife and young son.

They did not all leave together, however. Solon and his family travelled by rail to California, there to observe a solar eclipse, and on February 2, 1889, sailed from San Francisco down the coast to Panama. Marshall had the task of organizing all their equipment—some one hundred pieces of baggage—and would sail down the Atlantic seaboard to meet the others in Panama, whence they all journeyed together to Peru.

Solon kept a detailed journal of all that befell them. It was a slow journey down the Pacific coast, the ship calling at most ports en route, but this was as well, for the Baileys were very much newcomers to Latin America and had much to learn.

Mr Robinson, an American gentleman long resident in Guatemala and president of a railway in that country, gave me the following statement of affairs in the Central American Republics. Though called republics, they are in reality only despotisms of the most perfect type. In practice the president has nearly absolute power. Elections are but farces, no one venturing to oppose those in power. The only appeal from this order is by revolution. These are frequent, but usually futile. As in Mexico, the Indians are practically slaves, for by the laws a peon if in debt to his employer can be compelled to work it out. Nearly all the lower classes employed on the great plantations are in debt to their employer, and owing to low wages and careless habits are unable to free themselves from debt and at the same time from bondage. There is a middle class composed of those of mixed blood who despise the Indians, and in turn are despised by them; who have cut loose from the Church and have no faith at all. They are more quick and intelligent than the Indians, but also more dangerous, always ready for riot and insurrection. The people of these States are incapable of self-government. The condition of semi-slavery, though sad, is claimed to be necessary for the maintenance of any prosperity; some say the fault is in the lack of good government, but apparently the trouble lies deeper,—in the character of the people.

At Panama Solon and his family were met by "Mr. M. H. Bailey" and the hundred cases of equipment. The latter must have caused Marshall no small amount of effort, for the great canal was not yet finished, and everything would have had to have been off-loaded on the Caribbean coast, transported across the isthmus, and re-loaded on another ship at the Pacific coast. Indeed work on the canal had been almost abandoned, the De Lesseps Company being in receivership, and its labourers riddled with tropical disease. Solon took a gloomy view of things.

> We visited "La Boca", the mouth of the canal, during our stay in Panama. Work was practically suspended, however, and decay and desolation were the marked features. On the way we passed the company's hospital and burial grounds. The vastness of each bore witness to the power of the forces with which the promoters of the canal were obliged to contend. Panama is a city of considerable size, and owes its prosperity in recent years largely to the canal enterprise. A depression will doubtless follow as soon as this enterprise shall be definitely abandoned.

The ship bearing the re-united Bailey brigade lumbered on from port to port down the South American coast, Solon taking a rather disparaging view of most places. Finally, they reached Lima, where they spent two days arranging bank accounts and other business matters, before departing by train up the valley of the Rimac ("a small but tumultuous stream") for Chosica, the town chosen by Pickering as the centre of their explorations.

The moment they stepped off the train they knew that Chosica itself was out of the question as an observatory site; it lay in a narrow valley surrounded by bleak steep mountains which cut off much of the sky. So, while Mrs. Bailey and her son remained in a Chosica hotel to make frequent meteorological observations, the two Bailey brothers set off with a *cholo* guide to explore the surrounding area.

It was extremely hot and tiring work. Beyond the immediate vicinity of the Rimac there was no water or greenery anywhere, and the incredibly steep and high mountains meant that at one time they would be at dizzying altitudes and the next plunged into deep narrow valleys where the stagnant air was unbelievably hot. How precipitous the countryside is can be gauged from their finding themselves in a valley so narrow that their packmules were only just afforded a comfortable passage, while above them towered sheer 3,000-foot cliffs. Before long Solon realized that many of these valleys—clefts really—bore all the marks of dried-up river beds, the immense boulders strewn around bearing mute testimony to the power of the water that had once flowed in them. They soon heard of the extreme danger of being caught in one of these in the rare event of rain. The run-off from the precipitous mountains brought flash-floods of unbelievable rapidity. For example, near the village of San Bartolomé

is a deep railway cut terminating in a narrow valley, which is crossed at right angles by an iron bridge. The bed of the stream which this bridge crosses, usually dry, had been prepared for chance floods by enlarging it and paving the bottom near the railway with heavy smooth stones to facilitate the passage of bowlders and other materials brought down in time of flood. An eyewitness of the event, who was obliged to run for his life, says that the fall of water was so sudden that the mass descending the valley presented a solid wave-front twenty or more feet high. In five minutes the bridge was carried away, the natural outlet of the stream was blocked with stones and mud, and the torrent turned at right angles through the deep railway cut. In spite of this abrupt change of direction, within an hour the cut was filled for a distance of several hundred feet with a mass of mud and bowlders to a depth of ten, and in some places twenty feet. Some of the stones there deposited weighed several tons, and could not be removed by derricks without blasting.

Later, in the valley of the Verrugas, they saw a well-built railway viaduct that crossed the 250-foot-deep valley on strong pillars of iron and stone. Soon after, they were amazed to learn that the entire structure had been swept away in a flash-flood, and Solon would eventually find himself re-crossing the valley in a basket suspended from a single cable.

Eventually the Baileys made a tentative choice of a mountain only about eight miles from Chosica for their observatory.

This was our first visit to "Mount Harvard", as we called the hitherto nameless summit. It had been a hard climb, but we were repaid. Five miles away in a straight line a glimpse of green indicated the valley of the Rimac. The rest was hidden by mountains. In every direction nothing but barren mountains were to be seen. The soil was a hard

sand covered with huge bowlders and with many varieties of cacti.
To the north and south we looked down into gloomy ravines thousands of feet deep. It was nearing sunset, however, and the hoarse
whir of a condor's wings as he swept by over our heads admonished
us to be moving. Hastily returning to the spot where our animals
were fastened, we managed to get down to the valley by early evening. As we looked back from our hotel at the dark outline against
the sky, the thought of a residence on that isolated spot brought a
strange sense of gloom and loneliness.

The Baileys had now to return to Lima in order to build some political
fences; in particular, to pay their respects to the Peruvian President,
General Cáceres.

The audience was given in the *Casa del Gobierno*, which occupies
one entire side of the grand *plaza* of Lima. We entered through an
archway guarded by a sentry to the office of Señor Delgado. By him
we were conducted through a door usually kept locked, through an
inner court occupied by numerous soldiers, up a stairway and along
a corridor, both well guarded, to a locked door. After a slight delay
we were admitted into a large reception room, thence into an inner
reception room, and here the President met us. President Cáceres
was a tall, stern man, with battle-scarred countenance and erect,
soldierly bearing; his face showed the marks of privation and passion. Like so many of his predecessors in office, he had fought his
way to power, and the numerous soldiery and frequent rumours of
revolution gave evidence of the need for vigilance. The history of
Peru shows that the example of violence and bloodshed set by the
conquerors of the country has been followed too well, not only in
the times of the viceroys, but under the Republic also.

The President expressed a solicitous interest in their project, offered
whatever support he could, and insisted on presenting them with a letter
commanding all provincial governors to supply the Harvard expedition
with as many peons as might be needed for the building of an observatory or any other such purpose. Bailey, of course, politely accepted the
letter, but with his views on the peons being held in semi-slavery, never
made use of it.

Before settling on Mt. Harvard as their final choice, the Bailey brothers
decided they should make a sweep farther afield from Chosica.

About half-past eight in the morning of March 21 we left Chosica.
We had planned to start at five o'clock, and our mules were promised

for that hour; but it is seldom possible in Peru to move early or promptly. The road from Chosica to Matucana is pretty rough, and in many cases even dangerous. It is simply a mule path four feet wide, trailing along the face of the mountain, in many places lofty, steep, and slippery. Frequently it winds around the face of a cliff, cut into the solid rock, and the path worn smooth by innumerable feet for hundreds of years has nothing whatever to prevent the careless from slipping over the side and falling hundreds of feet to the river below. To add to the difficulty, one meets in a day hundreds of laden mules, donkeys, and llamas. The only safety is to insist on the inside track. Accidents are not rare. About noon we stopped at the little *pueblo* of Cocachacra for breakfast. We breakfasted in a little cane and adobe hut on *chupe,* eggs, and fowl. As the fowl was lean, tough, and half-raw, and the eggs few, our main reliance was placed on the *chupe.* This is a soup, and may be regarded as the national food of Peru. Briefly described, it is made of a little of everything the cook may ꞓ ᵼance to have in the house, but always with plenty of *picante,* or re⸍ peppers.

A short time after leaving Cocachacra we passed the Verrugas River, a small tributary of the Rimac. From this river, called *Agua de Verrugas,* is named the peculiar and dangerous disease known as Verrugas. At the time of the construction of the road it was especially prevalent in this vicinity, and from it arose a great mortality among the workmen. It is still common, and is found as far inland as Matucana. The disease is characterized by intense pain, but, especially, by the appearance on different parts of the body of sacs filled with blood. These are sometimes of a considerable size, and when the number of them is great, the loss of blood is considerable. Nevertheless, it is regarded as a favourable symptom to have the disease appear on the outside, and medicines are taken to produce this result. It was thought to be caused by drinking Verrugas water, and in this vicinity no one drinks water except in the form of *chicha,* the national drink.

As day succeeded day, they found themselves climbing ever higher in the cordillera, making for the town of Chicla, 12,220 feet above sea-level. Unaccustomed to high altitudes, they began to encounter the debilitating effects of mountain-sickness.

At the village of San Mateo, where we stopped for breakfast, one of the party, Mr. M. H. Bailey, suffered quite severely from *soroche* or mountain-sickness. It manifested itself by dizziness, faintness, nausea, and vomiting; complete unconsciousness occurred twice for a few minutes' duration. The patient was placed on the ground, and bruised garlic, the odor of which is thought by the natives to have great efficacy, was provided in abundance. A little hot soup, however, as soon as it could be provided, speedily brought the patient into better condition, and after a fair breakfast he was ready to proceed.

Solon would find himself no stranger to mountain-sickness either, and concluded that even the natives generally suffered from it above 14,000 feet, although the altitude at which one succumbed depended much on one's condition: those weary and exhausted from climbing fell prey far sooner than those who had arrived at a high altitude relatively relaxed by riding a mule.

They were having little luck in finding a site any better than Mt. Harvard, and much time was wasted when they made the grave mistake of explaining to local peasants what sort of place they were searching for and offering a reward to anyone who could take them to a likely site. This had the result of everyone knowing the perfect site, until, after the first few futile trips, they realized that for a tiny reward any peasant was prepared to lead them the length of the Andes in the hopes of coming on something that pleased the strangers. Accordingly, they returned to Chosica.

One advantage to Mt. Harvard was that it was within eight miles of the railhead at Chosica, so that bringing up their vast equipment would not be a superhuman task. Even so, the pathway to the mountain was in such a state of disrepair that three weeks were spent with a team of hired peons improving and widening parts of it. Despite this, some of the equipment on the heavily loaded mules was damaged as they wound through the rocky defiles. Later, when Bailey started sending back to Pickering some of the photographic plates that had been taken on Mt. Harvard, Pickering was rather startled at the damage to be seen on their edges. Had they been gnawed by insects, he asked. Replied Bailey,

> I am inclined to think that the plates have been injured near the edges by the tanks more than by insects. The only insects that appear to be abundant here are fleas and scorpions and they have thus far evinced a prejudice in favor of us rather than dry plates—of the latter we have found specimens in our hose, pantaloons, and coats but never any on the dry plate shelf.

One of the reasons why the Baileys had brought so much baggage was that they had been forewarned that Peru offered virtually no timber for building, so that a complete portable observatory and living quarters was part of their equipment. Getting it put up with local assistance proved an exasperating business.

The instrument building was also covered with paper in two layers,— black building paper outside and a strong manilla paper within. The natives employed to assist in this labor seemed unable to overcome their astonishment at this class of building, and insisted that adobe would be better. The fastening of the paper was quite laborious and attended with some ludicrous incidents; for our *cholo* assistant, more zealous than wise, was continually placing his ladder against the paper wall. The paper was strong enough to support him until he nearly reached the top, when he would plunge suddenly through the wall to his astonishment and my dismay.

With everything finally installed, the entire Bailey entourage plus several assistants settled down to life on the mountain and the real business of making astronomical observations. It was not a particularly easy life.

The mountain furnished neither water nor food. These were brought daily from Chosica eight miles away and nearly four thousand feet below. Considerable difficulty was at first experienced in finding a trustworthy man to act as muleteer. Those first obtained were drunken or lazy, so that we were repeatedly disappointed by the non-arrival of our day's water and provisions. Although we intended to keep several days' supply in advance, this was not always practicable, and several times we were for hours without water, and with none too luxurious a diet.

Animal life was by no means wanting in the vicinity. Of birds, the largest and those that most interested us were the condors. These monarchs of the Andes, with their immense spread of wing, paid us frequent visits. The sound of their wings, even when hundreds of feet above us, was peculiar and somewhat like the sound of wind blowing through the foliage of pines. When near it became a louder and fear-inspiring sound, hearing which, our domestic fowls hurried to some shelter, and even "Serrano" our dog seemed glad to retire to the safety of the house. Besides [the condors] there were a smaller species of vulture, and numerous eagles and owls. Occasional flocks of small green parrots were seen, always in considerable numbers and with a great deal of chattering.

Reptiles were frequently encountered near the dwelling, but they never became intrusive. Francisco, while chopping wood, occasion-

ally found one coiled up inside a dead tree. One day a snake with
fangs appeared in the yard near the house. Some hens were in the
vicinity, one with chickens. We were surprised to see one of these
hens walk cautiously up to the snake, and dexterously peck it in the
back close behind the head. The reptile was some three feet long and
apparently full grown, yet the hen managed herself so well that at
once the snake lay helpless, wriggling in the dust; whereupon Francisco despatched it. Scorpions were the most common pest on the
mountain, and might be said without much exaggeration to be
ubiquitous. They were found in all the buildings at times; one was
met with in a shoe, another in a coat sleeve, another in a bed, etc.;
yet in no case was any one stung, although there were several narrow
escapes.

Tarantulas were found, also, but only during the rainy season.
The largest of them measured seven inches stretch of legs. . . .

Life on Mount Harvard was somewhat lonely and monotonous.
Recreations were few and consisted chiefly of two kinds,—walks
about the mountains and lawn tennis. Usually the days were too
warm to render walking enjoyable during the middle of the day, but
early in the morning or about sunset these walks were very delightful.
The scenery in all directions was grand or beautiful, and the sun
setting into the Pacific often gave us superb sunsets. Our life was so
isolated that man and animal, dog, cat, and goat were on terms of
the greatest intimacy and equality. In our walks we were followed
at varying distances by the cat, the two dogs, and the two goats;
while even the fowls seemed to stroll in the same direction after us.
When we stopped to rest, dogs, cats, and goats would range themselves about, never going far away from us. No doubt this condition
of things was due to our lonely situation, and shows how all animals
feel the need of society.

For a different form of recreation we had cleared a small space
near the house of cacti and loose stones for a tennis court. Here, in
spite of many difficulties, we found a pleasant relaxation and healthful exercise. The fact that our balls sometimes rolled down the
mountain on one side or the other, so that they could not be found,
and that our court was very rough and uneven did not quench enthusiasm; and an hour was usually passed each day in this sport.


For a good many months everything went very well on Mt. Harvard,
and a steady stream of astronomical photographs and measurements
found its way back to a delighted Pickering. He wrote to emphasize the
importance of the work, and his pleasure at the way it was going. As
always he was solicitous of his distant team:

*The station on top of Mt. Harvard (by courtesy of Harvard College Observatory)*

I hope that you have taken all proper precautions against illness in the way of medicines and other supplies. Bad drinking water is always a source of danger . . . I am desirous that you should all be as comfortable as possible at your station, and hope that you will go to any reasonable expense to attain this end. The outdoor life and fresh air ought to be healthful and invigorating.

Pickering was never one to see anything incongruous about his sitting in Boston and offering detailed advice on how to conduct life in the Peruvian hinterland to those who were actually there.

But then came the rainy season, or more accurately, the cloudy season. Weeks became months when hardly any useful astronomical work could be done, and Solon began to realize that Mt. Harvard simply would not do as a permanent site for an observatory. He decided that he and Marshall would have to explore much farther afield than merely the area around Chosica, perhaps even going down into Chile and over to Bolivia. So, leaving a Peruvian assistant, whom Solon had been training, to make what observations he could from cloud-bound Mt. Harvard, the Baileys all departed for Lima. Here Solon's wife and son were to remain during the brothers' tour of several months.

Before starting, they were fortunate to encounter in Lima a British meteorologist, a Mr. Ralph Abercrombie, who had been making a meteorological study of the entire west coast of South America. "He spoke highly of Arequipa, Vincocaya, and Puno; but in a letter written later, he was of the opinion that the clearest point along the coast was somewhere in the Atacama desert [of Chile]."

The Baileys planned their travels around these suggestions. First they sailed down the southern coast of Peru, past Pisco (where "is exported the fiery drink called by the same name") to Mollendo, the nearest port to Arequipa and Puno. They found that the port "offers but little in the way of convenient landing of passengers or merchandise," ships anchoring a mile offshore, and passengers and goods being lowered in a cage to a waiting lighter, and then hauled ashore by crane. However, "Mr MacCord, formerly superintendent of the railway leading from this town to the interior, informed us that no passenger's life was ever lost, and no merchandise of much value."

Arequipa is some ninety miles or more inland from Mollendo, and these they covered by railroad.

At Cachendo, forty miles by the railway from Mollendo, we enter upon the pampa of Islay. From this point to Vitor, a distance of about forty miles, the road lies across the pampa,—a region void of vegetation and quite level except that it gradually rises towards the east. The ground is covered with rock and shifting sands, and all are heated intensely by the nearly vertical sun. Mirages are frequent; and the visions of cool and crystal water in the distance must have proved a tantalizing spectacle to the weary travellers afoot or mule-back in the ante-railroad days. Sand dunes in great numbers abound. . . .

As we approached Arequipa, however, fields of waving grain and groves of fruit-trees formed a delightful contrast to the region through which for several hours we had been riding. The first view of the city is really beautiful, surpassing in picturesqueness any other Peruvian city we had seen. It lies in the midst of a wide arable plain, stretching several miles in each direction. Above the city, which rests just at its foot, rises the "Volcano of Arequipa", El Misti, a nearly extinct volcano about nineteen thousand feet high. The city itself is built of white stone, which in the distance has the appearance of marble, and forms a pleasant contrast with the surrounding green fields. This stone, known as *sillar,* is a volcanic substance found in vast quantities on the flanks of El Misti, and is so soft that it is very easily worked. Indeed it is chopped into cubes with a sort of adze, much after the fashion of ice-cutting. The houses are almost universally of one story, and many of them show marks in ruined walls and other debris of the disastrous effects of the great earthquake of 1868. Before that date many of the buildings were of two stories; but at the time nearly all or quite all were thrown down, and now few persons care for more than one story. For the same reason the walls are built very thick, the walls of our room in the hotel being four feet in thickness.

Solon and Marshall paused only briefly in Arequipa before pressing on inland to Puno on the western shores of Lake Titicaca. In this region they found themselves consistently at altitudes of twelve to fifteen thousand feet, and the dreaded *soroche* was never far away. This, "together with the bleak, weather-scarred landscape, and scant vegetation," they found distinctly unappealing, and the little hotels they had to stay in hardly less so. Each room, Solon noted, contained a check-list of its entire contents, carefully gone over by the landlord when a guest was leaving. In most instances, "the list was not long enough to cause the landlord much trouble."

From Puno, as we saw, they sailed across Lake Titicaca into Bolivia.

*El Misti from Arequipa (by courtesy of Harvard College Observatory)*

The fuel used on this trip was llama dung, which is collected and sold by the Indians at about ten cents a sack. It makes a hot but not very even fire, as it is quickly consumed, and a large quantity is used on a single trip. In this connection it may be of interest to mention that on the trains from Arequipa to the interior a fuel is used called *yareta*. It is a species of moss which grows in dense, dome-shaped masses a foot or more in diameter and is decidedly resinous, so that it makes a very good and cheap substitute for coal.

Travel on the Bolivian plateau was slow, thirty-six miles requiring six hours, but the Baileys went across to La Paz.

"After Paris, La Paz", say the Bolivians, and we decided that it was a long way after. Still, with the brightly-colored roofs and walls of its buildings and the gayly-colored *ponchos* of the *cholos*, it presented a very brilliant and picturesque appearance.

But Solon had already made up his mind that the area was not suited to an astronomical observatory, and since it was already early December, 1889, and they had been gone almost a month already, he and Marshall made their way back as rapidly as possible to Arequipa. Rapidity was hard to come by, of course, on a 200-mile journey across rugged mountains towards the coast. Day succeeded day. Nights in primitive adobe structures, days spent in little rickety railroad cars, the Indians riding atop the roof, chattering and uncaring of the stupendous scenery. The train wound a crazy course, looping around the precipitous mountains, and the occasional wash-outs where the track navigated water-courses necessitated long delays.

When they finally reached Arequipa, an event occurred that probably set the entire future course of Harvard's southern observatory: Solon got sick. Thus instead of spending only a day or two there, the brothers spent two weeks in Arequipa, and while Solon languished in bed "Mr M. H. Bailey kept a careful record of the cloudiness." They became very impressed with the clarity of the skies, the natives assuring them that this was indeed the start of the rainy season and that the present weather was not at all unusual for that period. This almost certainly influenced the ultimate choice of Arequipa as the permanent site for the observatory, although time would show that the natives had been wrong: those two weeks were distinctly better than average for the rainy season.

Meanwhile the brothers went back to the coast and started a long and tedious journey down the coast of Chile. They went as far as Valparaiso and inland to Santiago, where they visited the National Observatory.

> Though possessed of a considerable equipment, the instruments seemed either to be unmounted or out of repair, and evidently the Government has not shown a keen enough interest in its prosperity. Señor Obrecht of the Observatory informed me that he had passed several months at Copiapó [a coastal city to the north on the Atacama Desert] and that that city and region offered better conditions for astronomical work than the neighborhood of Santiago.

So now they began a slow journey northwards again, stopping at most of the little coastal ports such as Coquimbo, Copiapó, and Antofagasta, and hiking inland away from the cold clammy coastal fog to investigate sites. Soon it became clear that Señor Obrecht had been right; the desert foothills of the Andes offered incredibly good astronomical conditions. A month was spent at a place called Pampa Central in the heart of the Atacama, where they set up various instruments to check the skies. Out of twenty-nine nights twenty-eight proved to be excellent; they later learnt that at Mt. Harvard there had been only two good nights in the same period.

But if the Atacama was an astronomical paradise, it was utterly repellent in almost all other respects. The desolation was appalling. Marshall noted that in one run of 140 miles they saw no green thing, not even a cactus. Everywhere across the harsh landscape were nitrates, and indeed, almost the only Chileans there were miners working the nitrate diggings and the silver mines. At times water was unobtainable. It had to be scooped from 200-foot wells, and one relied on *cholos* who brought it great distances across the desert in bags by donkey-back. It had, said Solon, a very unpleasant taste, but was reputedly safe.

So it was that when Solon made his final report to Pickering, although he could see the astronomical advantages of the Atacama, the pleasant white houses and green surroundings of Arequipa exerted their pull. For Pickering, always solicitous of human conditions, that was enough. The Harvard station would be in Arequipa.

But that was still some way off. In March of 1890 the Bailey family returned to Mt. Harvard to find a depressing state of affairs. Not only had very little astronomical work been accomplished, but the torrential rains had quickly shown up the inadequacies of paper-clad houses. A fair amount of flood damage was done before Bailey got the paper replaced with heavily painted canvas. There they all remained for several more months, while Solon wrote his reports and Pickering made plans and decisions at Harvard. Finally, in October of 1890 there began the dismantling and transport of the equipment from Mt. Harvard to Arequipa. The first part of the journey was the worst.

> The whole equipment of the station reached Chosica in safety, although several mules, made unsteady by loads of lumber, rolled down the mountainside for some distance. No bones were broken, however, and no special damage done. All the instruments were either carried by hand or on the backs of mules that were led by hand.

And now Edward Pickering felt that the Baileys had done enough. In what would prove to be a disastrous decision, Edward sent out his brother William to take over affairs and choose the actual site near Arequipa. The Baileys were invited to take a leisurely trip home via Cape Horn and Europe before Solon started work at Harvard again.

William Pickering arrived with his family and two assistants on January 17, 1891, and there was several months overlap with the Baileys so that he could hear first-hand reports and a precise site could be established. It was not until May that the Baileys started for home, and the journey was not quite as relaxing as they had expected.

> The journey along the west coast [of Chile] was relieved from monotony by incidents in the Chilian civil war. All ports north of Coquimbo were in possession of the congressional party, while those to the south were still held by President Balmaceda. War-ships of both parties chased each other up and down the coast, and took and re-took the slightly defended ports. Our steamship, though sailing under the British flag, which commands respect in Chili as elsewhere, was, nevertheless, delayed in entering and leaving ports. Our captain did not think it safe to enter after dark, lest we should be mistaken by some war vessel for an opponent, or damaged by torpedoes. At Iquique we were pleased to see the American flag flying from two handsome war vessels, the "Baltimore" and "San Francisco", and on the following morning from still another, the venerable "Tallapoosa". To the Itata incident was due this unusual display of the stars and stripes. We anchored near the "Cochrane", a heavy and slow war-ship of the congressional party. Just at dark a

cannonading was begun by some invisible vessels two or three miles
out to sea. It was said to be the "Imperial", a fast merchant steamer
that had been taken by the President and fitted with some guns,
accompanied probably by two small but swift torpedo boats recently
received from Europe. These three vessels constituted at this time
the main naval strength of the President. On this occasion it seemed
to be pure bravado, as no one could see what object was to be
gained by such a demonstration. All stood on deck watching the
flashes and listening for the reports. Whatever its meaning might be,
our neighbor, the "Cochrane", evidently did not approve of it; for
considerable activity was visible aboard her. The anchors were
hoisted, and she steamed close by us out to sea, and soon dis-
appeared in the darkness. Her departure had good effect, for the
firing ceased after a time. In the morning the "Cochrane" was back
in her old place, and the sailors were cleaning their guns. Shortly
after her departure the night before, a terrific explosion startled us
and shook our steamer. Women screamed, and the first impression
was that an attempt had been made to blow us up. We had been in
no such danger, however. A small torpedo boat had been passing,
carrying a torpedo said to have been made by amateurs. When near
us the torpedo was accidentally exploded. The torpedo boat was
completely demolished, and the three or four men aboard were
killed. At Caldera we saw the wreck of the war-ship "Blanco". A
short time before our arrival the captain of this vessel, flushed with
some victory, went ashore to attend a banquet; while those on board,
apparently unsuspicious of danger, exercised little caution. Nearly
all were undoubtedly asleep when one of the President's torpedo
boats, already referred to, crept near the "Blanco" and exploded two
or three torpedoes under her. She sank almost immediately with her
crew of about two hundred men. At the time of our visit the bodies
of one hundred and twenty victims were still in the wreck. The
"Blanco" went down in shallow water, so that some of the spars
projected above the surface; and a little beneath it in pleasant
weather, a ghastly picture presented itself to the eyes of the curious.
At several ports we were detained for some time, while search was
made by the authorities for a political refugee, who was thought to
have escaped from Iquique and to be on board. In spite of the most
careful searching he was not found; but when we arrived at Coquimbo, a port in possession of the President, he came forth from his
hiding-place, with one arm in a sling and with other marks of suffer-
ing and privation. He was arrayed by his friends in full uniform,
and paraded the deck with evident pride. One report was circulated
that he had been concealed in a large cask and fed through the
bung-hole by a sailor bribed for the purpose, and another that he had
been concealed in a seaman's chest.

At Valparaiso were the war-ships of several nations, and the merchant ships of many. The guns of the fort covered the harbor, and an electric search-light swept the surface of the water by night to detect any suspicious arrival. Our captain, whose naturally hasty temper had been sadly disturbed by the indignities he had undergone during the voyage, expressed the wish that under some pretext the English war-ships would open fire on the fort, declaring that one of them, the "Warspite", would be amply sufficient to demolish it. Of the truth of this statement we had no means of judging, as that highly interesting experiment was not tried. After refusing to receive us for half a day, as punishment for having called at rebel ports, the authorities at Valparaiso finally relented, and we were allowed to board the "Galicia", our English steamer bound for Europe. Thereafter we had no more forcible reminder of war than the presence of a courteous Chilian gentleman and his wife, who were suspected of sympathizing with the congressional party, and who consequently found it unsafe to remain in the country. The voyage in the South Pacific was rather boisterous; but . . . all on board were cheerful. We reached Bordeaux July 4, and Cambridge [Massachusetts] on August 15, 1891.

William Henry Pickering was a disaster. Unable to take direction himself, he was equally unable to give it without arousing enormous antagonisms in others. He was arrogant, aloof, and had an overriding sense of self-importance that completely blinded him to the most elementary concepts of commonsense. As an astronomer he was, if anything, worse.

It is strange that Edward Pickering, the very antithesis of his brother, should have been so blind to these faults as to have invited William to give up his post as a physics instructor at the Massachusetts Institute of Technology and join the staff of the Harvard College Observatory in 1887. True, Edward made him just one more assistant on the staff, but William immediately assumed that his fraternal status automatically made him second-in-command of the observatory, and acted accordingly, Now, in Peru, he would very nearly bring total disaster to his brother's carefully made plans and Solon Bailey's two years of exploration.

He arrived in Arequipa with his wife, two children, his mother-in-law, a nurse, and two assistants (rather carefully chosen, because William did "not care especially for a brilliant man"). Edward's plan had been for his brother to lease land on which to erect the observatory, and he had rather naively assumed that the Pickering entourage would simply take over the large house already rented by the Baileys when they had brought the equipment from Mt. Harvard. Edward had been quite explicit over the leasing of land: William was not to spend more than $500 on it.

The first intimation of things to come arrived in the form of a cable from Peru saying simply—William was not one to waste words—"Send four thousand more." The bemused Edward could hardly believe his eyes. "Account overdrawn. Economy necessary. Name minimum. State object," he replied, but this only produced an even more insistent demand from his brother. Fearing some major disaster had overtaken William, Edward cabled the four thousand dollars with a stern warning: "Use care. Money scarce. Write fully."

When a letter did finally arrive from William, Edward must have been aghast at what he read. William had bought a large piece of land outright, had begun construction of a large house complete with servants' quarters and a stable, had erected lavish buildings to house the telescopes, and was ordering a good road to be built to the site from the nearby town. His letter concluded with thanks for the $4000 and a request for another $2000 to cover running expenses and $7000 for the building of the house.

What was Edward to do? If he refused additional money, that which had already been sent would have been wasted, yet the interest earned by the Boyden Fund could not possibly sustain this kind of spending. After a series of what must have been exceedingly disagreeable meetings with the University's administration, Edward wrote to say that they had been allowed to borrow money against the Fund's capital, and implored his brother to keep spending to an absolute minimum. William's response was an announcement that he had hired another local assistant who would need paying, and the suggestion that another telescope be sent down with funds for the construction of its housing.

So much for administration. On the scientific front things were even more embarrassing. Edward had a very detailed long-term plan for the work of the southern observatory. It would complement the work done at Harvard, and would be mainly photographic, involving a large-scale astrophysical survey of many southern stars. Now, presumably hardly recovered from the financial nightmare over costs, Edward sat back and awaited the early arrival of the photographic plates by which he hoped to begin justifying the recent expenditures. To no avail. William wrote to say he had five hundred plates and would send them on soon. Edward, with some acerbity, replied "I am very glad you have 500 plates but very

The Arequipa station photographed on February 17, 1892, with William Pickering's residence as the centrepiece (by courtesy of Harvard College Observatory)

sorry they are not here." Another nine months without plates finally drove Edward's patience to the limit. "Photograph with thirteen inch [telescope]," he ordered by cable. But William was happily engaged on something entirely different.

The close opposition of Mars to the earth in 1877 had been the occasion for many astronomers around the world to make a detailed study of the planet. One of them had been David Gill on Ascension, but the fame that Gill had acquired from that work was exceeded by the fame that the event brought to Giovanni Schiaparelli, the forty-two-year-old director of the Brera Observatory in Milan. Schiaparelli seized the opportunity of Mars being at its closest to the earth to study the surface details visible on the planet. He thought he could discern a series of faint lines across the face of Mars, which he described in Italian as "canali." The English translation of this word is best given as "channels," but it is hardly surprising that it was initially translated as "canals." There is a distinct difference: "channels" could refer to naturally occurring features—as Schiaparelli intended—but "canals" implied an artificial origin. If there were canals on Mars then someone must have built them, and so began decades of violent arguments over whether or not Mars is inhabited and just what the Martians were up to with their canals.

Now to give William Pickering his due, he did have astonishingly acute eyesight. Where most people can see about half-a-dozen stars in the Pleiades, William could see twenty-three. To one with his eyesight, the problem of whether or not there exist canals on Mars must have been particularly appealing, so that when the planet again made a fairly close approach to the earth, in August 1892, Pickering was hard at work with his eye glued to the telescope trained on Mars, heedless of Edward's cries that instead he photograph stars.

It was very soon evident that William's eyesight was exceeded only by his imagination. He saw no reason to restrict the reporting of his results to the usual channels of the scientific literature, and began firing off exciting cables direct to the *New York Herald*. The first intimation Edward had of what his brother was doing astronomically, was a stunning report in that newspaper that William had discovered great mountain ranges on Mars, that the polar icecaps of Mars were melting near their edges to form rivers flowing towards the Martian equator, and that he had seen at least forty lakes varying in size from 80 by 100 miles to 40 by 40 miles.

The astronomers of the Lick Observatory were reported to have received this intelligence "with a kind of amazement." Working under
equally good conditions and with a telescope three times the size of
William's, they were quite unable to see any of these things.

Edward, already beginning to hear sniggers among the astronomical
community, rushed for pen and paper to set William straight.

> The telegram to the N.Y. Herald has given you a colossal newspaper
> reputation. A flood of cuttings have appeared, forty nine coming this
> morning. In my own case I should have restricted myself more dis-
> tinctly to the facts in this as in other cases. You would have rendered
> yourself less liable to criticism if you had stated that your interpre-
> tations were probable instead of implying that they were certain.

But William was off and running. Having disposed of Mars he turned
his attention to the four great moons of Jupiter. The result was another
immediate cable to the *Herald*: "The first satellite is egg-shaped and re-
volves end over end, and nearly in the orbital plane. Its period is twelve
hours and fifty-five minutes."

This was too much for the Lick astronomers, among them experts in
this particular area. There must have been some hopeless mix-up in the
transmission of William's cables. "Very likely the telegrams are wrong?"
the Lick Director enquired hopefully of Edward. Unfortunately, they
were not wrong; it was William's imagination getting the better of his
eyesight.

Even Edward could see the writing on the wall. If the Harvard College
Observatory was not to be irreparably damaged both in its reputation
and in its finances, there could be only one course of action. William must
come home and the reliable Solon Bailey must be sent to replace him and
get on with the proper work of the observatory. Edward wrote a tactful
letter explaining this to William, and to ease any upset he suggested that
William might like to stay on briefly to do his visual work, or even return
to Peru in the future while Bailey carried on Edward's programs. But he
took care to make it plain that Bailey would now be running the station,
and "would probably wish to occupy the present house."

William was outraged on all counts.

> Without being boastful, I think I've accomplished a pretty big thing,
> and if the authorities could see it they would say I had got them a

great deal for their money. Now it turns out that the running expenses will be about equal to the annual income [of the Boyden Fund], or have been so the past six months. Perhaps they may be less in future. I shall make them as small as possible, but we must live, and live decently. . . . Now what are you going to do about it? That's what I want to know. . . .

As to our coming down here again to Peru and living in a small hut, while the Baileys occupy the Director's house, it is out of the question. I planned and built that house, and while I am in Peru I expect to live in it. I don't choose to live in a shanty while one of my subordinates occupies the house I built.

I like Peru, and I like to live and work here, but when I am here this is *my* Observatory, as a department of the Harvard College Observatory, and it is not Bailey's Observatory nor anybody else's. If Bailey and I are here together again, I want him to distinctly understand . . . that as you said to me when I first came to Cambridge, that under you, *I* am at the head of this Department. I have no equal in authority, and only one superior, and that is yourself. If Bailey and I are here together again, that is to be understood, and he must understand that he is only at the head of this Observatory because I am away. This Observatory is not to be run by a Council while I am in it. . . . As you like Cambridge and have a talent for publicity, and as I like Peru and have I think a talent for overcoming photographic and mechanical difficulties, I think we can run this establishment very well between us.

Edward's reply was bleak. Yet it announced his readiness to make a personal financial sacrifice to help his brother.

The last of [your letters] was not a pleasant one to receive either as your superior officer or as your brother. In future semi-official communications of this kind please use forms of expression which cannot be used to your disadvantage if seen by others. The President and Fellows of Harvard College are not easily bullied and had your letter been presented to them the results might have been disastrous to you. As soon as possible after learning the unfortunate financial position of your account with Grace and Co., I sent all the facts to President Eliot [of Harvard], with a portion of your letter, omitting all the objectionable portions. With regard to the deficit you have incurred you write "Now what are you going to do about it? That's what I want to know." Under the letter of the Bursar, the [Harvard] Corporation does not appear to be responsible for this deficit. If this view is insisted on, "what I am going to do" is to help you out of this difficulty by paying a portion of the deficit myself. I do not think the Corporation will insist. . . .

Your position as regards the Boyden Fund appears to need explanation. When I spoke to you about coming to Cambridge I told you that your position would not be an independent one. Your case is like that of Mr Wendell who has charge of the 15 inch equatorial. In each case you carry out my plans and I hold you responsible for their execution. . . . The Corporation having given me absolute authority in these matters, it is not my intention or policy to delegate it to another. If you suppose that I gave you more authority than this, it arose from a misunderstanding and I now withdraw it.

William began to see that bullying was not going to make his brother change his mind, and now both William and his wife wrote lengthy letters to Edward pleading their case to stay in Peru, offering regrets that more photographic work had not been done, and promising to make amends in the future if they could only stay on. But Edward was implacable; they and their assistants must return forthwith. Bailey was on his way.

Solon, his wife and son and two fresh assistants arrived in Arequipa on February 25, 1893, to find the truculent William, his family, his assistants, and several guests, still occupying the observatory house. The Baileys were forced to stay in a hotel for a month until the Pickerings felt ready to leave. Relations during the transition period must have been cool indeed.

William Pickering was never reformed. Years later he managed to convince Edward that Harvard should have an observatory near Mandeville in Jamaica, with, of course, himself in charge. It was the same story all over again. William did little of the work that Edward assigned him, and instead turned to visual observations of the moon. It must have been little of a surprise to the long-suffering Edward when there came a sudden inundation of reports that the moon had canals, snowcaps, a dense atmosphere, and vegetation. Reported William enthusiastically:

Whatever reputation as an astronomer I lost when I published my former observations, will be nothing to the destruction produced when these get into print, & especially the drawings. I have seen everything practically except the selenites themselves running round with spades to turn off the water into other channels!

In 1924, five years after Edward's death, the Harvard College Observatory cut William and his Mandeville Observatory adrift, but William stayed on there, maintaining the observatory as his own, absorbed and unrepentent, until his death in 1938.

Meanwhile Solon Bailey in Arequipa was looking about him with no little despair. William's tenure had left the station in a sad state of disrepair. One telescope was "encumbered with old iron and stones . . . no possible use for this rubbish could be found," others were broken and had parts missing, the clocks were unreliable and frequently stopped, everything was out of adjustment (the earthquake recorders had been so set up that they recorded an earthquake every time the door was opened), and some of the meteorological equipment was still in the packing cases as it had arrived from Harvard.

The accounts of the Observatory were even more depressing:

> Heretofore, to judge by present customs, servants have been accustomed to obey everyone—assistants have ordered in the city articles according to their wishes without control either for the Observatory or for other purposes. Assistants (& servants?) have been accustomed in town to draw money without authority from the grocer and have it charged against them. Private bills of all sorts have been run up by the assistants and the accounts settled by the Observatory.

It took Bailey a great deal of hard work and the instigation of some very firm new rules ("these radical changes seem to cause some friction with the old assistants. . . .") before he could at last settle down to Edward's long-delayed astronomical programs.

There were other things to do also. During one of the prolonged cloudy spells (which Bailey gloomily noticed were more frequent than he had ever expected), he tried his hand at a little archaeology. With the help of a guide he did some excavating of an Inca burial ground in the vicinity of Arequipa. Although others before him had worked the area, he had the good luck to find an untouched grave almost at once.

> The grave contained one skeleton and three pots or jars. Curiously I hit upon this grave the first time and Francisco [the guide] was so impressed with my skill that he lifted his hat to me four times while digging and thought I must have done it by Astronomy.

Meteorology was a more serious preoccupation. Edward Pickering was enthused with the idea of establishing a network of reliable weather stations in the area, and Solon resolved to have one on the very top of 19,000-foot El Misti.

He had previously tried climbing the Misti and had had to give up when overcome by mountain-sickness. Still, William Pickering had succeeded, and indeed it had been climbed as early as a hundred years before, when a party of priests had struggled to the top and erected a large iron cross there, a cross which in Bailey's day was viewed by the Arequipans as the special protector of the city. And Bailey's previous experiences in Peru had shown him the way in which the *soroche* was to be overcome, or at least minimized: one must keep fresh for the highest altitudes by going as high as possible on muleback. This Bailey went to considerable lengths to achieve. Parties of Indians were sent to hack out a track for the mules, and a hut was constructed as a way station for the climbers at the 16,000-foot level. In late September of 1893 Bailey and one of his assistants, George Waterbury, led the assault on the mountain. The careful preparations proved their worth, and the party arrived at the summit in relatively good condition. Whereupon the Indians

> all embraced me in turn and afterward embraced G.A.W. Then each removed his hat and kissed first the base and then the arm of the cross. Then they drank our health and afterward or before dug a little hole and each placed therein a little coca and poured on it a little wine and covered it over. . . . A little of the romance of mountain climbing is perhaps lost by using mules legs, in part, instead of depending entirely on ones own: but what is lost to sentiment is many times made up to science, for one arrives at the summit with quiet nerves and in good condition for exact work. I think I may claim to have taken mules higher than they ever went before in any well authenticated instance. Perhaps this may prove to be a step in an evolutionary process, from the primary dependence on personal endurance to the time when scientists will regularly ascend to great heights for meteorological study, by means of captive balloons or flying machines. I hope our mules on our next expedition will appreciate their honorable position as connecting links, and exert themselves accordingly.

Solon, who eventually died at the age of 75 in 1931, would live to see the beginnings of his remarkable prophecy come true.

So successful were these innovations in mountain climbing that a weather station with self-recording instruments was readily established on El Misti's summit, and the collecting of the data became a routine in the lives of Bailey's assistants. The Bishop of Arequipa asked if he might send his secretary along on one trip to celebrate High Mass on the summit

*An ascent of El Misti—a photograph taken at about 18,000 feet on January 5, 1894 (by courtesy of Harvard College Observatory)*

*The shelter on top of El Misti, October 12, 1893. Standing at right foreground is Solon Bailey ( by courtesy of Harvard College Observatory)*

(and probably also check that the astronomers were holding the cross in due respect). Reported Bailey to Pickering,

> I think I shall go with him myself, in order to be sure that he takes no offense at anything. Our station is rather close to the cross and a trifle higher but I think this padre is a reasonable man and will find no fault.
>
> I imagine the ceremony up there will be interesting and I am sure it will be the "Highest" Mass ever celebrated. When he gets us properly blessed, I think he would be willing to go to Cambridge for a reasonable price and finish the [Harvard] Observatory.

But such humour soon began to give way to more ominous reports. By late 1893 the political situation in Peru was becoming tense, and before long it was clear that another revolution would be forthcoming. At first Solon took the possibility quite lightly, and jocularly wrote to Pickering that in the event of war he might "have to remove the lenses and use the telescope tubes for cannon." Within a few months, however, his reports had a much more serious tone to them, and he asked for authority to buy heavy shutters to protect the house.

> We have two or three revolvers and with the addition of a few good clubs, I think we should be able to keep off any drunken rabble, who would probably be the only serious source of danger. If a really serious armed party demanded the surrender of the house we should have to "hands up" and rely on the government for indemnity. We shall also lay in a special supply of provisions as at such times no marketing can be done. I do not really expect any trouble but yet I think it well to be on the safe side. The extra expense will probably be in the vicinity of $100.00 and the shutters for the windows and doors will possibly be useful in later years, unless the Peruvians learn wisdom, and the food we can eat in any case.

Throughout 1894 the alarums of war steadily intensified, and towards the end of the year mounted troops of the rebel leader Nicolás de Piérola began overrunning the smaller towns of the country. Riots broke out in Arequipa, and several people were shot. Back in Cambridge the alarmed Pickering pulled strings sufficient to get the Secretary of State to cable the American Minister in Lima "It is feared Arequipa Observatory endangered by present war. Spare no effort for its protection." But there was little the American Minister could do except call on the already beleaguered Presidential forces for protection, and they were by then fighting for their own existence.

The Baileys had their first real taste of revolution in January of 1895, when they were returning from Mollendo by train and found themselves captured by rebels.

> I was reading [reported Solon to Pickering], when we heard a tremendous shout of "Viva Pierola". I looked out only to see a crowd of men armed with rifles and revolvers come rushing around the train and into the car. The car was at once filled with cries of "Jesus Maria" and "Por Dios", by the ladies and children. The natives really seemed much more frightened than we were. I advised Mrs Bailey and Irving to keep quiet and there would be no harm done and so it turned out. The revolutionists behaved with great moderation and offered us no indignity whatever. We were sent back to Mollendo however while the men followed us in another train which they had captured. When near the town they left us locked in the car and forming in line marched in and took the place in minutes. Mollendo is said to have a population of about 3000 but there were only 15 soldiers and they surrendered after about a hundred shots were fired. After the affair was over we were allowed to enter the town and found everything quiet.

It did not stay that way for long. Government troops arrived to recapture the town, and the Baileys spent an anxious night in the home of a Mr Golding, agent of the steamboat lines. Mr Golding and Solon stood guard through the night armed respectively with a revolver and club. But there was little bloodshed in the government troops' victory, and the next morning the train passengers were once again dispatched on their way to Arequipa.

Within a fortnight of this incident, the war came to Arequipa in earnest when the town was attacked by rebel forces. Bailey decided the time had come to put their emergency plans into operation. He removed the fragile telescope lenses from their tubes and buried them in a deep pit, noting laconically that it was as well that the revolution had come during the cloudy season. Then they closed their new shutters over the windows and doors of the house, ran up the American flag, and sat back to await the end of a couple of weeks of fighting. Work was not entirely neglected, though, for each evening one of the assistants would more matter-of-factly than heroically emerge to read the meteorological instruments while firing continued just beyond the garden.

In Cambridge the normally mild Pickering was raised to heights of warlike passion. As usual he wrote off reams of practical advice to his man on the spot, noting comfortably that the main danger would likely come

from attacks of drunken or disorderly stragglers rather than from regular troops. Perhaps it would be as well, in view of future disorders, to put the Observatory or at least a portion of it in a condition in which it could safely be defended from stragglers. A stone house which cannot be burned from the outside is not readily captured by a mob. The greatest danger is that using a stick of timber as a battering ram, any ordinary door can be broken in. The scarcity of wood in Peru diminishes this danger. Loop holes through which you can see what is going on, and if necessary fire without danger might prove very useful. During the troubles in Ireland a form of country house was proposed as shown in the annexed sketch. The circular bay windows in the corners have steel shutters with loopholes commanding all four sides of the house. It was claimed that four determined men, armed with repeating rifles, could hold such a house against a mob of any size, unless it was provided with artillery. . . .
I have an improvement to suggest to your proposed line of action in case of an attack, but I wish that your position might be behind a steel lined loophole. Can you conveniently pour buckets of water (preferably hot) on persons attempting to force an entrance. The most dangerous mobs are sometimes dispersed by a thunder shower.

Perhaps fortunately, this warlike advice only reached Bailey after the battle was all over. The revolutionaries won the war, and Piérola became President. The Baileys thought it politic to throw a garden party for the new administrators, Solon apologizing to Pickering for the additional twenty-dollar expense so incurred.

Thereafter the Arequipa Observatory settled down to long years of solid astronomical work. Edward Pickering's vision did not fail him, for the work done there would in time form a cornerstone of twentieth-century astronomy. This work was nothing less than the first large-scale astrophysical survey of the sky. Whereas previous star catalogues had concentrated on listing little more than the accurate positions of the stars, the Harvard catalogues gave data about the stars themselves: their brightnesses and spectra, for instance. To this day they are an indispensable tool of the astronomer. Edward, in fact, has never received the full recognition that is his due from historians of astronomy: without his foresight and determination, brought to fruition by Bailey, twentieth-century astronomy would have been much diminished.

The cloudy season at Arequipa, however, became an increasing irritation, and more and more Pickering began to think of a change in site. The Atacama Desert seemed as inhospitable as ever, but Pickering's old friend David Gill kept reiterating his suggestion that South Africa offered good astronomical sites. So it was that in 1908 Solon Bailey, now in his fifties, found himself setting off again on yet another tour of exploration on yet another continent. He travelled widely through the Cape Colony, the Orange Free State, the Transvaal, and Rhodesia, before choosing a site near the town of Bloemfontein in the Orange Free State. But the intended move was long delayed by another financial crisis at the Harvard Observatory, and it was not until 1927 that Harvard finally closed its doors in Peru and left to continue the work in South Africa.

It is pleasant to be able to say that Solon Bailey was as good an astronomer as he was an explorer and administrator. During his years in Arequipa he chose as his own special field of study the beautiful globular star clusters, the most prominent of which are seen in the southern skies. In these he discovered a particularly interesting kind of variable star, and to this day astronomy students are well-accustomed to speaking of the Bailey-type classifications of these stars, even if hardly anyone recalls anything of the man himself.

It was perhaps Edward Pickering who years before, in another context, chose a phrase that best sums up the contrast between Solon Bailey and William Pickering, both as men and as astronomers. "Science," said Edward, "is an enobling pursuit only when it is unselfish." No less, he might have added, is that true of life itself.

How much the Whisper, how much the Vision, how much the Need? What were the motives that drove these astronomers on their marches across the world? One might argue that Charles Mason or Thomas Maclear or Solon Bailey had no particular motives, that they were merely obeying the orders of higher authorities who told them to go, but that would be to beg the question. Any one of them could have found a way out if the prospect of their travels had seemed too appalling. And certainly John Herschel or David Gill were under no compulsion to set out across the oceans.

Was it some personal need? Were these men in some way psychological misfits who, like Lawrence of Arabia or "Chinese" Gordon or Richard Burton, were driven by deep inner compulsions to escape ordinary life into extraordinary worlds of their own? Almost certainly not. A few, like William Pickering, may have been mildly neurotic, but the great majority were eminently sensible, friendly, family-loving people. Solon Bailey, a wise and kindly administrator; David Gill, fun-loving, offering brandy and cigars to his youthful future king and exchanging Scotch jokes with the Governor-General of Canada; John Herschel, the sophisticated and urbane dinner-companion of Mayfair hostesses. And even after his many years away, Le Gentil was accused of no more than a touch of brusqueness. Social misfits they were not.

Were they visionaries? The answer must surely be yes, although they were visionary in a pragmatic, perhaps even prosaic way, and not in an ecstatic, rapturous sense. For astronomers in general, paradoxical though it may seem, are among the least starry-eyed of people. They are forever wary of those who speak of their overwhelming fascination with the universe, of their amazement at the scale of space and time, of their feelings at being such a speck in the universe. Astronomy requires more intellectual discipline than that. Yet it would be wrong to suppose, as Walt Whitman once did in a well-known poem, that astronomers never see or feel beyond their mathematical equations. There *is* a mystique to astronomy, but it is a subtle mystique that comes only through a deep understanding of the subject, the kind of mystique that makes great art and great music endure.

Added to that are other motives, including the sense of achieving some degree of immortality, however miniscule, by adding to man's understanding of the universe. But the ultimate motive remains enigmatic. Stars are studied for the same simple reason that mountains are climbed: they are there. As a former student once said of Arthur Eddington, one of the most famous of twentieth-century astronomers: "He was the sort of fellow who might have gone out and shaken his fist at the stars and declared 'I'm going to find out about you!' "

So, yes, our travelling astronomers were visionaries. They understood that without their efforts astronomy must be delayed. Why else would Le Gentil have waited out eight years abroad, or Gill have given over half a year to living in desolation? Yet their visions never overwhelmed their pragmatism. What more practical than Le Gentil seizing control of his ship in the China Sea or Chappe hunting down his guides in the Siberian forests, Green chasing thieves across the beaches of Tahiti, or Bailey burying his lenses to the sound of gunfire?

But vision must have its trigger. There must be the Whisper of some little voice that challenges ordinary, unromantic, everyday people to seize the opportunity. For Gill the little voice that said Why Not? when the Cape Observatory was unexpectedly offered to him, for Herschel the sudden realization that his father's work would remain incomplete without a survey of the southern sky. Without the Whisper the Vision will never be seen.

And what of the modern astronomer? Undoubtedly there are more astronomers travelling the world than ever before. The problems of convincing distant administrators of one's needs in the field remain, but little else. Where once the peripatetic astronomer dealt in shipwreck and war and mule-back ascents of the cordillera, his modern counterpart is more concerned with the quality of airline food, and whether Monday's flight out of Santiago will get her to Geneva in time for Tuesday's committee meeting. For the twentieth century is an age of transition. It has seen the end of a millenium that opened up the world to travel.

But with the twenty-first century will come a new age, the era of exploration beyond the earth. Immense voyages to strange planets will be undertaken, and the face of astronomy will be utterly changed, both by the surprises that will emerge by visiting those planets, and, for the astronomy of stars and galaxies beyond the planets, by giant new telescopes floating weightless in the still calm of interplanetary space. A new generation of peripatetic astronomers will come, and, as always, among them will be those ordinary yet remarkable people who will know the vision and hear the whisper. Especially the whisper.

SELECTED BIBLIOGRAPHY

1

H. Woolf, *The Transits of Venus,* Princeton University Press, 1959.

H. S. Hogg, *The Journal of the Royal Astronomical Society of Canada,* vol. 42, pp 153-159, 189-193, 1948.

H. S. Hogg, *The Journal of the Royal Astronomical Society of Canada,* vol. 45, pp 37-44, 89-92, 127-134, 173-178, 1951.

J. C. Beaglehole, *The Endeavour Journal of Joseph Banks,* Angus & Robertson, London, 1952.

J. C. Beaglehole, *The Life of Captain James Cook,* Stanford University Press, 1974.

2

D. S. Evans, *Science,* vol. 127, pp 935-948, 1958.

D. S. Evans, T. J. Deeming, B. H. Evans & S. Goldfarb, *Herschel at the Cape,* University of Texas Press, 1969.

*The Sky,* vol. 1, Nos. 4, 5, 6, 1937. (Reprinting of the Great Moon Hoax.)

3

I. Gill, *Six Months in Ascension,* John Murray, London, 1878.

G. Forbes, *David Gill: Man and Astronomer,* John Murray, London, 1916.

D. Gill, *History and Description of the Cape Observatory,* His Majesty's Stationery Office, London, 1913.

4

S. Bailey, *Annals of the Harvard College Observatory,* vol. 34, pp 1-48, 1895.

B. Z. Jones & L. G. Boyd, *The Harvard College Observatory,* Harvard University Press, 1971.